PALLADIUS

THE WORK OF FARMING

PALLADIUS

THE WORK OF FARMING
(Opus Agriculturae)

and

POEM ON GRAFTING

A new translation from the Latin

by

JOHN G. FITCH

PROSPECT BOOKS

2013

First published in 2013 by Prospect Books,
Allaleigh House, Blackawton, Totnes, Devon TQ9 7DL.

© 2013, English translation and editorial matter, John G. Fitch.
© 2013, illustrations, Chris Mundigler, International Contract Archaeological Research Services.

BRITISH LIBRARY CATALOGUING IN PUBLICATION DATA:
A catalogue entry of this book is available from the British Library.

Typeset and designed by Lemuel Dix and Tom Jaine.
The cover illustration is of a mosaic at the Gallo-Roman villa at Seviac, Montréal (Gers), France. Photograph by Tom Jaine.

ISBN 978-1-903018-92-7

Printed and bound by the Gutenberg Press, Malta.

Table of Contents

Acknowledgements

My thanks are due to my colleagues at Victoria, John Oleson and Geof Kron, for discussion and support; to Tom Jaine for readily undertaking the publication of this translation; to Chris Mundigler for preparing the drawings; and to Palladius himself for providing a congenial task that engaged me over some difficult years. I am much indebted to previous translators, particularly to C.F. Saboureux de la Bonnetrie, Thomas Owen, René Martin and John Henderson. Anyone who works closely on the text of the *Opus Agriculturae* stands on the shoulders of two giants of Palladian scholarship, Josef Svennung and Robert Rodgers.

John Fitch
Victoria, B.C.

Acknowledgements

Introduction

Palladius

His full name, according to the manuscripts, was Palladius Rutilius Taurus Aemilianus; the quadruple form indicates that he came from a family of high standing, and the names Palladius and Rutilius point to an origin in Gaul. We know little about him beyond what he tells us himself, namely that he had farms on Sardinia and in Italy near Rome. His date is much debated, but he probably wrote in the fourth or fifth century AD.[1] The manuscripts call him *vir inlustris*, a title which began to be employed in the second half of the fourth century, initially for men of the highest rank in the Roman Senate, and later more widely. If correct, this title gives a *terminus post quem* of *c.* 370 for the publication of his work, and at the same time indicates his high status.

By his own account, Palladius had considerable personal knowledge of farming. There are some 19 places where he refers explicitly to his own experience, mostly concerning fruit trees, and at least seven further places where he states his opinion in such a way as to imply experience.[2] He seems to have had a special interest in fruit trees, which occupy more pages in the *Work* than either vines or vegetable gardens, and are also the subject of the *Poem on Grafting*.

1. Sardinia: 4.10.16, 12.15.3. Italy: 4.10.24. Near Rome: 3.25.1. His references to farming experience in 'cold' or 'very cold' regions (4.10.15, 8.3.2) suggest somewhere other than Sardinia or Rome. For full discussions of his date, see Martin 1976: vii–xvi, who is inclined to place the *Opus Agriculturae* as late as 460–480 AD, and Bartoldus 2012: 11–36, who dates Palladius' life *c.* 400–470, and the *Opus* before 455.
2. Citation of experience: *2.13.8*, 2.15.1, *3.18.6*, 3.25.20, 3.25.22, 3.25.27, 3.25.31, *3.26.5*, 4.10.15, 4.10.16, 4.10.24, 8.3.1, 8.3.2, 11.12.5, 12.7.1, 12.7.8, 12.7.12, 12.7.22, *12.15.3* (the few passages *not* concerned with tree fruits are italicized). Opinion implying experience: 1.28.5, 2.9.1, 3.10.4, 3.24.8, 6.2.1, 11.8.2, 14.27.1 (none on tree fruits).

Despite his status and the wealth that must have accompanied it, Palladius presents himself as a down-to-earth farmer who appreciates simple practical methods. In the farmhouse itself, he commends ceilings based on wooden planks or mats of cane, because those materials are readily available on the farm. For propagating olive trees he recounts not only the labour-intensive method recommended by his predecessor Columella (3.18.6), but also the 'easier and more practical' method which he knows is used by most farmers. Similarly, where Columella had given elaborate instructions on keeping chickens, Palladius in his brief discussion envisages a more typical farmyard flock, which largely looks after itself. On raising asparagus, since growing from seed is a slow and painstaking process, Palladius prefers the 'cheap and thrifty' method of transplanting wild asparagus roots, or else the convenience of purchasing ready-grown clusters of crowns; neither shortcut is mentioned by Columella.[3]

Tapping the tradition

In reading Palladius and his predecessors, we are reminded that they belong to a long tradition of organic farming. While it is true that they farmed organically out of necessity, in the absence of artificial fertilizers and pesticides, it is also true that they did so productively and successfully. Columella had laid down one of the basic principles: that when forested land is converted to farmland, the soil is initially fertile because it contains a large amount of organic matter, but as that material is used up, the farmer needs to replenish the soil's fertility (2.1.5–7). In Palladius we find a variety of methods still used by organic farmers today: rotation of crops, fallowing, growing 'green manure' crops to be ploughed-in, fertilizing with manure and urine and ash, amending the texture of the soil with material brought in from elsewhere. We also recognize that elegant frugality of organic farming, in which nothing produced on the farm is wasted: even the water in which beans have been cooked has its value (for strengthening peach trees, 12.7.4).

3.   Ceilings: Palladius 1.13.1. Olive trees: Columella 5.9.1–4, Palladius 3.18.6. Chickens: Columella 8.2–7, Palladius 1.27. Asparagus: Columella 11.3.43–46, Palladius 3.24.8, 4.9.10–12.

Reading Palladius and his predecessors also reminds us that their farming reflects the Mediterranean diet, defined by a substantial intake of olive oil and wine, a heavy reliance on vegetables and fruit, and a relatively low consumption of meat. In antiquity olive oil was essential not only for cooking, but also for lighting and as soap. Wine was drunk regularly by all classes, though usually heavily diluted with water. The importance of oil and wine is reflected in the number of recipes provided by Palladius for seasoning them. He also testifies to a great variety of fruits, not only to staples such as apple and pear but also to exotics such as jujube and sebesten. In the absence of refrigeration, preserving fruit was a challenge, and Palladius devotes considerable space to discussing methods of preservation.

In addition to his own farming experience, Palladius draws chiefly on three authorities. For field crops, including vines and olives, and for animal husbandry he relies on Columella, who wrote in the first century AD. As his chief source on vegetable gardens and fruit trees, Palladius uses Gargilius Martialis, who wrote in the third century; since these works of Martialis are now largely lost, Palladius gives us valuable indications of their content. Third, for more exotic material such as recipes for flavoured wines, Palladius draws on a compilation of agricultural information by the fourth-century Greek writer Anatolius of Beirut.[4] A detailed listing of Palladius' sources, paragraph by paragraph, is provided in Rodgers' edition of his text.

Palladius' achievement was to draw this abundance of information into a clear well-organized handbook of manageable length for the practical farmer. One means by which he did so is to abbreviate the sometimes wordy discussions of his predecessors, often with a notable gain in clarity and cogency. As an illustration one could compare Palladius' discussion of peach trees with that of his source Gargilius Martialis.[5] Martialis adopts a leisurely pace suited to a specialist treatise,

4. Anatolius' compilation does not survive in its original form: later Greek writers continued to add to it, until it was finally codified in the tenth century AD in the form in which we now have it, the *Geoponika*. For the sections on farm buildings in Book 1, Palladius makes use of the *Synopsis of Private Architecture* published by Cetius Faventinus in the third or fourth century AD.
5. Palladius: 12.7.1–8. Martialis: Mai 1828: 394–403, or Condorelli 1978: 18–27.

citing the views of multiple authorities, often in disagreement with each other, and taking 700 words to cover the topics of seeding and transplanting alone. Palladius by contrast is businesslike and incisive, using less than one-third as many words for the same topics, and writing with an authority born of experience.[6]

Another means by which Palladius abbreviates is to excise material of marginal relevance. Columella, in his desire to cover every aspect of agriculture, had included highly specialized topics such as breeding wild animals (oryx, wild boar and the like) and fish-farming of marine species; Palladius omits these. Again, Columella had provided voluminous lists of varieties of vines and fruits, but Palladius takes the sensible view that it is pointless to list varieties, since the farmer should rely on those that have been tried and tested in his locality. No doubt he took to heart the danger ruefully acknowledged by Columella himself, that an exhaustive, encyclopaedic treatment is likely to overwhelm the farmer and scare him off (1 pref. 28).

Palladius' greatest innovation, however, was to reorganize his source material on a calendrical basis, with a book devoted to each month of the year. While there were various brief precedents for a farmer's calendar,[7] no one so far as we know had organized a whole treatise in this way. Whether Palladius worked chiefly from memory of his sources, or used a written filing system, the task of rearrangement was complex and substantial. Sometimes a single paragraph of Palladius pulls together material that was scattered in three or four places in the sources. The benefit to practical farmers of the calendrical plan is obvious: they can simply turn to the relevant book to see what tasks need to be undertaken in any given month. Each month is organized on the same pattern, which makes it easy to find one's way around: first field work including olives and vines, then vegetable gardens, then fruit trees, next livestock, and finally miscellaneous topics. Palladius increased the ease of use by adding, at

6. It must be acknowledged that Palladius' abbreviations are sometimes too drastic. In particular, he often omits the reason for taking some action, leaving the farmer puzzled.
7. They include Hesiod, *Works and Days* 383–616; Varro, *On Agriculture* 1.29–37; Columella 11.2; Pliny, *Natural History* 18.230–320; and Book 3 of the *Geoponika*. An inscriptional calendar (*CIL* VI 1.2305) is reprinted in abbreviated form by White 1970: 194–95, with comparisons to Palladius.

the front of each book, a summary of its contents. The goal of all his changes, then, was to produce a manual that would be accessible to his readers, by the restrained scope of its contents and its convenient format.

As with content, so with style. The brevity of the preface to the work, which occupies just 14 lines of Latin, exemplifies Palladius' aim of accessibility through succinctness. At the same time, it announces a goal of stylistic accessibility. Palladius protests that most instructors inappropriately address plain farmers in a highly sophisticated style, to the point of incomprehensibility.[8] This implies that he himself will use a simple, comprehensible style, as indeed he does. For the most part, his sentences throughout the work are relatively short, with unambitious syntactical constructions. He makes much use of second-person verbs, which give directness and vividness. Columella had written, 'Now other crops that can be worked when wet are nevertheless better hoed when dry, because when handled in this way they are not attacked by rust; barley, however, must not been touched unless perfectly dry' (2.11.5). Palladius writes, 'If you hoe the crops when dry, you have given them some help against the rust as well. Barley in particular should be hoed when dry' (2.9.2). The gain in vigour is evident, and the second-person verbs establish a relationship between writer and reader, acknowledging the involvement of the farmer in the action. Similarly this relationship and involvement is suggested by the use of first-plural verbs. On making honey, for example: 'But before we press them, we shall cut away the parts of the combs that are spoilt or contain the young, since their bad flavour spoils the honey' (7.7.3). Here again Columella had used passive forms, with less directness: 'But care must be taken that those parts of the waxen cells that contain either the young or dirty red matter are separated from them; for they have a bad flavour and spoil the honey with their juice' (9.15.12).

---

8. Is this critique aimed at Columella? The length of his seven-page preface certainly contrasts with the fourteen lines taken by Palladius. But while Columella could be accused of being prolix and orotund, he is hardly incomprehensible, except perhaps by exhausting the reader's attention. Given the loss of other agricultural manuals, we cannot say whether Columella is the chief target.

Nevertheless, like any educated writer in antiquity, Palladius is a conscious stylist, taking care, for example, that his sentences end with rhythmically satisfying clausulae. One important criterion for the ancients was *variatio*. In introducing the 'hours' at the end of each Book, for example, Palladius scrupulously varies the wording: 'This month [January] tallies with December in the length of the hours'; 'For determining the hours, this month [March] agrees with October'; 'In the measurement of hours, May corresponds to August.' He also observes lexical variation (e.g. *truncus* and *lignum* for 'tree trunk' at 3.25.6) to avoid excessive repetition of the same words.[9] Occasionally he indulges in a liking for balance between phrases or clauses, usually to end a paragraph with a flourish, e.g. 'If the branches are burdened by a heavy crop of apples, all the flawed fruit should be thinned out, so the sap can share nourishment equally among the rest, *and provide the well-bred apples with the abundance that was being wasted on a multitude of mean ones*' (*et generosis abundantiam ministret quam numerosa uilitate perdebat*, 3.25.16). Such stylistic flourishes are associated particularly with the topic of 'uniting diverse elements,' as in grafting. For example, he recommends grafting the cultivated olive onto wild olive stock below ground level, and concludes as follows: 'From the cultivated olive it will retain a fertile capacity to regrow above ground; underground, from its bond with the wild olive, it will retain a fruitful ability to thrive' (*quae et apertam redeundi felicitatem de olea et occultam ualendi feracitatem de oleastri conexione retinebit*, 5.2.2).

The silences of Palladius

Besides the omissions mentioned above, Palladius is silent about two topics that had engaged Columella: finance and manpower. The primary motive for these omissions was presumably to keep the length of his manual within bounds. Their effect, however, is to place greater

9. For another example of variation on his predecessors' phrasing, see the comparison of Palladius and Martialis on almond trees by Martin 1976: 198–99. In my translation I have generally avoided using different terms with the same reference (i.e. in a case like *truncus* and *lignum*) to avoid confusion.

value on the procedures of farming themselves, and to free the farmer to decide for himself about finance and manpower, in accordance with his circumstances.

Columella's stated goal had been to show that farming (as opposed to business, law, or politics) was a viable means of 'safeguarding and increasing' inherited wealth (1 pref. 7–10), by which he clearly meant substantial wealth appropriate to the élite of the Roman empire. In keeping with this goal, his eye was regularly on profit (*quaestus*), and he began his discussion of several branches of farming by assessing their profitability. By removing financial calculations, Palladius implicitly shifts the emphasis from profit to productivity. Where Columella says that rich loose-textured soil produces the greatest profits (*quaestus*, 2.2.5), Palladius says that such soil produces the greatest yield (*fructum*, 1.5.6). Correspondingly Palladius envisages the farmer as engaging in his tasks 'for the sake of pleasure and productivity' (*ratione uoluptatis et fructus*, 1.1.2), which suggests that farming is undertaken for values intrinsic to it, rather than for the sake of financial profits resulting from it. Furthermore, by removing his advice from the context of a particular social class, Palladius gives it wider applicability, and allows the farmer to make his own decisions about the relative importance of marketing and self-sufficiency.[10]

Palladius also says very little about manpower. Columella had given detailed calculations, for example about the number of man-days required to produce arable crops. Palladius eliminates most of these calculations: he takes the sensible view that 'the calculation of how many workers are required cannot be uniform, since lands are so diverse; familiarity with the soil and region will readily show what numbers should perform any given task' (1.6.3). Since he assumed a regime based primarily on slave labour (though with possible use of tenant farmers also), Columella had discussed the housing, treatment and duties of slaves, both chained and unchained, and devoted two books to the

---

10. Palladius' only references to growing for market are 1.26 (thrushes) and 3.26.2 (piglets), both from Columella. Elsewhere he mentions selling only what is not worth keeping (8.4.5, 11.14.3, 12.13.8, 14.33.2). His sole references to buying-in are 1.6.4 (straw), 1.38 (bees), 4.9.11 (asparagus 'sponges') and 4.11.1 (cattle).

role of the overseer (himself a slave) and his wife. By contrast Palladius makes only passing references (once each) to agricultural slaves and tenants. He does use the non-specific term 'workers' (*operae* or *operarii*) somewhat more often, but not in such a way as to suggest a particular organization of labour.[11] It appears that he is deliberately avoiding references to a particular social system, in order to be as flexible and widely applicable as possible.

The relationship of Palladius' manual to the agricultural and social conditions of the late empire has been much debated, but inconclusively, for various reasons. First, there are varying views on conditions in the countryside at this period (and conditions certainly varied from one province and region to another). Second, as we have just seen, Palladius deliberately eschews those references to social organization which would be of interest to historians. Third, most of what he does say is inherited from earlier writers, which makes it hazardous to draw specific conclusions from it about the conditions of Palladius' own day.[12]

Correspondingly it is difficult to say that Palladius' work envisages a particular type of farm. He gives directions for growing olives and vines, but does not insist that farmers must grow them. Similarly he does not imply that a single farmer will raise all the species of fowl that he lists in Book 1, or all the varieties of fruit trees that he discusses throughout the work. One might perhaps infer from his work a preference for a moderate-sized farm in which the owner is personally involved in operations and keeps an eye on the health of the workers (14.1). But this is more like an ideal or archetypal farm than a specific system. It could easily be adapted to other situations, e.g. a farm run by a *procurator*, or an estate belonging to a church.

There are other silences in Palladius too. Columella had devoted space to such topics as 'whether the earth is growing old' and 'the

11. Columella on slaves: 1.6–9; on overseer and wife: Books 11–12. Slaves in Palladius: 1.6.18 (the other references, at 1.9.4 and in the preface to the *Poem*, are to domestic slaves). Tenants: 14.29.4. *Operae* or *operarii*: 1.6.3, 3.9.13, 6.4.1, 7.2.1, 8.1.1, 9.3, 12.14.
12. Attempts to relate Palladius to late-antique agrarian conditions include Kaltenstadler 1984, Morgenstern 1989 (remarkably uncritical), and Grey 2007. More skeptical is Vera 1999.

mythical origin of bees', though he acknowledged that the latter topic is better suited to leisured students of literature than to busy farmers.[13] Palladius evidently agrees with his assessment, and omits such material. More striking still, Palladius steers clear of the affective and ideological associations of farming. He has none of the city-dweller's longing for a supposedly healthier or simpler way of life, none of the antiquarian nostalgia found in Vergil's *Georgics* and in much English writing about the countryside, and equally none of the military connotations of farming, embodied for Romans in the legend of Cincinnatus. He writes for practical farmers who are less interested in fine ideas than in producing fine crops.

Book 14

In the *Poem on Grafting,* Palladius refers to the 'twice seven books' of his *opus agricolare,* his farming work. Readers were puzzled for centuries by the reference, since only Books 1–13 existed in the main manuscript tradition. The puzzle was not solved until in 1925, when Josef Svennung discovered Book 14 nestled between Book 13 and the *Poem* in a Milan manuscript. Book 14 is concerned with veterinary medicine, a topic not treated in the first thirteen books.

Palladius prefaces Book 14 as follows: 'So there should be no omissions in this work, I have collected together the medical treatments for all kinds of livestock and farm animals, and taken care to lay them out in a single book ... using the very words of Columella and his sources.' It looks, then, as if the addition of this book was an afterthought. In Books 1–13 Palladius had systematically omitted veterinary treatments, which form an integral part of Columella's discussions of livestock in his Books 6–7; no doubt his motive was, once again, to limit the length of his manual. But later, perhaps as a result of readers' criticisms ('so there should be no omissions'), he decided to add the veterinary material. The fact that he chose to copy Columella's very words, rather than scrupulously revising and rephrasing as he had in Books 1–13, suggests that he was now keen to finish the work quickly. Even here,

---

13. Preface to Book 2, and 9.2.

however, he remembered his practical purpose, by compiling a list of the *materia medica* that the farmer-reader should keep on hand in case of emergencies, and prefacing the book with it.

The Poem on Grafting

Palladius' account of the genesis of the poem is given in his prefatory remarks to his friend Pasiphilus: he was so embarrassed by his scribe's slowness in preparing a presentation copy of the *Opus agriculturae* for Pasiphilus, that he added the poem by way of compensation. In some sense, however, the poem had long been waiting to be written, since, as we have seen, the topic of 'uniting diverse elements' (as in grafting) had regularly inspired Palladius to shift into a higher stylistic register in the *Opus* itself. It is also reasonable to recognize a further level of motivation: the poem is a demonstration, to Pasiphilus and other *literati*, that Palladius is capable of writing in a more cultivated manner than the spare, businesslike style which he adopted for the *Opus*.

The poem, then, is not an integral part of the *Opus*: it is directed to a different audience, and virtually all its information on grafting had already been given in Books 1–13. In this respect it differs from Columella's verse Book 10 on gardening, which was part and parcel of his whole work on agriculture. In another respect, however, Palladius' poem is akin to Columella 10, in that both belong to an immensely long tradition of poetry of knowledge – a tradition which was already ancient when Hesiod composed his *Works and Days* in the eighth century BC and which inspired major works of Latin verse, Lucretius' *On the Nature of Things* and Vergil's *Georgics*. In the context of this genre, Palladius made an intriguing decision, namely to write in elegiac couplets (hexameters alternating with pentameters), rather than the standard metre of continuous dactylic hexameters.[14] It is tempting to wonder if this choice signalled that the poem was a jeu d'esprit, a literary exercise, rather than a serious exposition. But since the elegiac couplet was a favourite

14. The traditional equivalent for the elegiac couplet in English is the rhyming 'heroic couplet'. But I found the demands of rhyme too constricting to the task of conveying Palladius' meaning, and therefore chose the more flexible medium of unrhymed iambic pentameters.

medium of poets in the late empire, used for a wide variety of subjects, we should probably not read too much into Palladius' choice.

Influence and reception

The earliest reference to Palladius' work is by the statesman and scholar Cassiodorus in the sixth century. He records that he has deposited manuscripts of Columella and Palladius, and of writings on gardening by Martialis, in the library of his monastery at Squillace, to assist the monks in growing food for themselves, for pilgrims and for the poor and sick. Cassiodorus praises Palladius for his 'very clear elucidation' of his subject, whereas he describes Columella as providing 'a most satisfying banquet' of information, but as 'more suitable for the learned than for the untaught' (*Institutions* 1.28.6).

In the centuries that followed, Martialis' work on gardening and fruit trees ceased to be copied anywhere in Europe, and is now known only from meagre fragments. Columella disappeared behind the walls of a few monasteries, until his text was discovered at Fulda by Poggio in 1418. Palladius, however, by virtue of his brevity and accessibility, continued to be used, and carried forward the knowledge of Roman agriculture into the Middle Ages and the early Renaissance.

About 100 Latin manuscripts of Books 1–13 of Palladius have survived the hazards of use, worms and fire into the present day. The earliest are six copies made in the ninth century, during the Carolingian revival. They were all written in France, and apparently all copied, at first or second hand, from the same exemplar. In the twelfth century the text reached England, and in the thirteenth century it was known in Italy and Spain.[15]

Some of these manuscripts, including several of those copied in England in the twelfth century, are elaborate productions, not intended

15. R.H. Rodgers, *An Introduction to Palladius*, London 1975, is (despite its title) a detailed study of the manuscripts and textual tradition. See also Rodgers in P.O. Kristeller (ed.), *Catalogus Translationum et Commentariorum* Vol. 3 (Washington, D.C. 1976) 195–99, with useful older bibliography. For the reception of Palladius and the other Roman agronomists, especially as revealed by readers' annotations, the indispensable study is Mauro Ambrosoli, *The Wild and the Sown: Botany and Agriculture in Western Europe 1350–1850*, Cambridge 1997.

for everyday use. In other copies, however, readers' annotations and underlinings bear witness to the fact that Palladius was being read actively, with an interest in his content. Some of the Carolingian copies, for example, contain glosses and marginalia, particularly on topics that might be relevant to northern Europe, such as bee-keeping. A copy made in England *c*. 1300 has parallel passages from Vergil's *Georgics* added in the margins. In the sixteenth century an anonymous English reader compiled a Latin-English glossary of botanical terms to assist his understanding of the text. In the same century Sir John Prise made an alphabetical index of topics on the end-papers of his copy, and annotated certain sections of the text, notably Book 1 on building, at about the time that he was constructing a new stately home in Herefordshire.[16]

Naturally Palladius appealed *inter alios* to those with an interest in gardening. In the ninth century the monk Walafrid Strabo, known for his poem *Hortulus* on his herb garden, copied Palladius' recommendations on peach and pine trees (12.7.1–12) into his notebook. Five centuries later Petrarch made occasional jottings about his gardens in the back of his copy of Palladius.[17] These jottings show that Petrarch stood somewhat obliquely to tradition, and liked to experiment: for example, at Parma in 1349 he had vines planted at the foot of trees in Roman style, but they were fruit trees, not the standard Roman support trees. But his attitude is not entirely foreign to the spirit of Palladius, who describes his own trials with the grafting and growing of fruit trees, and who writes of a certain Greek recommendation *opus est experiri*, 'tests are needed' (3.29.3).

Translations and adaptations of Palladius

A sure sign of the perceived usefulness of Palladius is the very early appearance of vernacular translations. The first Italian translation, in the Florentine dialect, appeared in 1350; a second translation, in the Umbrian dialect, was printed in 1526, and a third, by the enterprising

16. Ambrosoli 1997: 16, 36–38.
17. Translation and commentary by William Ellis-Rees, 'Gardening in the Age of Humanism: Petrarch's Journal,' *Garden History* 23 (1995) 10–28.

polymath and printer Francesco Sansovino, followed in 1560. In what
is now Spain, a version in Catalan was composed in the second half
of the fourteenth century; it was soon followed by a Castilian version,
based on the Catalan and commissioned by Peter of Aragon. In France
the first translation did not appear until 1554; but this was only because
a Palladius-derived work by Crescenzi, discussed below, had received a
French translation in 1373, which cornered the market for decades.[18]

England lagged somewhat behind the Continent; nevertheless
the date of the first translation, c. 1442, is very early indeed.[19] It was
commissioned by Humfrey, Duke of Gloucester, son, brother and uncle
of kings; this was about the time that Duke Humfrey made a rich gift
of manuscripts to Oxford University, which necessitated the building of
what is still known as Duke Humfrey's library (part of the Bodleian). The
translator is thought to have been Thomas Norton, chaplain to Duke
Humfrey and chancellor of his household. The commission reflects the
Duke's interest in humanistic learning, and it seems to have been his
decision that the medium should be verse, no doubt to confer a certain
literary cachet on the translation. The form chosen was the rhyme royal,
introduced by Chaucer in the previous century, consisting of seven-line
stanzas in iambic pentameter with the rhyme scheme ababbcc. Norton
(if it was he) handled this potentially restrictive form skilfully, with
variation of rhythm and pauses; in general he translated accurately,
though the fixed rhyme scheme sometimes constrained his ability to
convey Palladius' meaning clearly. The following sample, chosen almost
at random, is taken from the vegetable garden in February.

Now cunula is sowe & hath culture
As oynons or garlec; & now cerfoyl
After this monys ydus do thy cure

18. Ambrosoli 1997: 17–24.
19. There are two editions: B. Lodge (ed.), *Palladius on Husbondrie*, Early English Text
Society 52 and 72, London 1873 and 1879; M. Liddell (ed.), *The Middle-English Translation
of Palladius De Re Rustica*, Part 1, Berlin 1896 (only Part 1 published). Liddell's is the better
edition from a superior MS, but Lodge provides useful if brief notes. The work deserves
a detailed edition that would elucidate its botanical, agronomic, historical and literary
aspects.

To sowe in faat & moyst, ydonged soyl.
Now betes sowe, & synke or quaterfoyl
Transplaunte; and somer thorgh hem may me sowe,
In faat land, moyst & donged wol they growe.

The rootes wold in donge ydippid be,
And delve hem ofte and make hem feeste of donge.[20]

Not only translations but also adaptations of Palladius' work reveal
his influence. Both the great encyclopaedists of the Scholastic age,
Albertus Magnus and Vincent of Beauvais, use him extensively. Vincent
does so throughout his *Speculum Naturale*, particularly in Books 5–6
and 9–22. In Book 5, for example, which is on the waters of the earth,
he quotes Palladius 9 on detecting and testing water, on digging wells
and aqueducts, and later Palladius 1 on cisterns, fishponds and baths.
Albertus' use of Palladius, on the other hand, is largely concentrated in
one section of his *Historia naturalis*: Part 18 is concerned with vegetative
plants (*de Vegetabilibus*), and itself contains seven books, of which Book
7 is devoted to agriculture; it is section 2 of this, containing lists of field
crops, herbs, vegetables and fruit trees, including vines, that follows
Palladius closely. Vincent's method is to copy out passages verbatim
from his authorities, citing them by name; Albertus, by contrast,
rewrites Palladius' material in his own limpid style, though he does
occasionally refer to Palladius by name in section 1 of Book 7.

These huge encyclopaedias were hardly convenient for practical use.
It was a different case, however, with the best-known medieval handbook

20. Book 3, lines 652–58. Palladius' original runs thus (3.24.9–10): *Cunela etiam nunc seritur
et colitur eo more quo alium uel cepulla. nunc cerefolium locis frigidis post Idus seratur; desiderat
agrum laetum, umidum, stercoratum. Hoc mense betam seremus, quamuis possit et tota aestate
seminari. Amat agrum putrem, umidum locum. Transferenda est quattuor aut quique foliorum
radicibus fimo recenti oblitis. Amat frequenter effodi et multo stercore saturari.*
'Summer savory is sown now as well, and cultivated in the same way as garlic or onion.
Now, after the Ides, chervil should be sown in cold places; it wants land that is fertile,
moist, manured. This month we shall sow beets, though they can be sown throughout the
summer. They love crumbly land and a moist location. They are ready for transplanting
(roots smeared with fresh dung) when four- or five-leaved. They love being dug around
frequently and drenched with lots of manure.'

of farming, the *Ruralia Commoda* of Pietro de' Crescenzi, completed some time between 1304 and 1309.[21] After a career in law and politics, which included 25 years of travel in northern Italy, Crescenzi had retired to his estate near Bologna to farm and write. In compiling his manual he used Palladius as the primary source of information, generally copying him out verbatim rather than paraphrasing; he did, however, reorganize Palladius' material from a month-by-month system to a topic-based system (e.g. Book 5 on trees, Book 6 on vegetable gardens), thus unwittingly reversing Palladius' own reorganization of Columella. To supplement Palladius he added material from Varro, particularly on livestock and poultry. There are also citations of Cato, and of Book 4 of Vergil's *Georgics* on bee-keeping. Crescenzi did not, however, have access to texts of Martialis or Columella, and his citations of them are taken from Palladius. Avicenna was used for the medicinal virtues of plants and other health-related issues, and Crescenzi drew on more recent authorities including Albertus Magnus (notably for pleasure gardens in Book 8) and Giordano Ruffo (especially for horse medicine in Book 9). He also added observations based on his travels and his own farming experience.[22]

This combination of ancient authority with modern additions proved irresistible. The number of surviving manuscripts of Crescenzi's work is extraordinarily large – at least 143, of which some 100 have the original Latin text, while the remainder have vernacular translations.[23] An Italian translation appeared in 1350, and contained an added chapter on rice growing, then being introduced to Italy; a French version, or rather adaptation, was commissioned by Charles V of France, and appeared in 1373. The advent of printing led to a publication explosion: 15 separate printed editions of Crescenzi's Latin text came onto the

21. Will Richter (ed.), *Petrus de Crescentiis: Ruralia Commoda*, 4 vols., Heidelberg 1995–2002.
22. On Crescenzi's sources see Richter (previous footnote) 1.xxv-lxxiii.
23. 132 manuscripts were listed by L. Frati in *Pier de' Crescenzi (1233–1321): Studi e documenti*, ed. T. Alfonsi et al., Società Agraria di Bologna, Bologna 1933, 259–306. Nine more are added by Robert G. Calkins, 'Piero de' Crescenzi and the Medieval Garden', in Elisabeth B. MacDougall (ed.), *Medieval Gardens*, Washington D.C. 1986, 155–73, at p. 159 fn.9. To these may be added Prague, National Library MS. V E 26 (Ambrosoli 43 fn.4) and New York, Pierpont Morgan Library, MS. B.17. 'Undoubtedly there are others which have not yet been publicly recorded' (Calkins).

market between 1471 and 1548, and a further 42 editions in various vernacular translations in the period 1478–1583.[24]

For obvious reasons of cultural and geographical distance, Crescenzi's work never achieved the same degree of popularity in England, though at least 19 manuscripts survive there of the Latin text. But an English translation was never made. When King Edward IV wanted luxury copies of Crescenzi, he ordered two of the French version, and one of the Italian, from Bruges. And it is in an Italian copy that Edward VI left a poignant memorial – neat underlinings in the text, and a drawing of an oak-tree with acorns, made in the last years of his short life.[25]

Another adaptation of Palladius, much shorter than Crescenzi, circulated widely in the fifteenth and sixteenth centuries. This was the curious *Tractatus Godefridi supra Palladium*, attributed to Gottfried von Franken. Organized on a month-by-month basis, it concerns itself with the growing and grafting of fruit trees, the cultivation of the vineyard and storage of wine. This little work did receive an English translation, variously entitled *Geoffrey upon Palady* or *Godfrey upon Palladie*, which survives in some 20 manuscript copies. It seems to have appealed to an interest in the exotic, and in such topics as growing stoneless fruit, or producing novel fruits by grafting.[26]

Printing immensely increased the circulation not only of the various translations and adaptations of Palladius, but also of his own Latin text: the 71 years between 1472 and 1543 saw (depending on how one counts) some 20 separate printed editions.[27] By the mid-sixteenth century, however, farming manuals were being produced in the vernaculars by writers familiar with local conditions. Palladius' centuries of renown

24. These figures are taken from Calkins 160. Ambrosoli (1997: 48) estimates that 12,500 printed copies of Crescenzi were put on the market from 1471 to 1564.

25. Nineteen MSS: John H. Harvey, 'The First English Garden Book,' *Garden History* 13 (1985) 83–101, at 100 fn.7. Edward IV and VI: Ambrosoli 84, 90.

26. Ambrosoli 29–34. The Latin text was edited by W.L. Braekman, *Geoffrey of Franconia's Book of Trees and Wine*, Brussels 1989; the English translation by D.G. Cylkowski in L. Matheson (ed.), *Popular and Practical Science of Medieval England*, Michigan 1994, pp. 301–330.

27. Conveniently listed in Rodgers' edition of Palladius, p. xxiii. All of these editions contain not only Palladius but also the other *scriptores rei rusticae*, viz. Cato, Varro and the rediscovered Columella.

were over, and he was to fall into ever-deepening obscurity. But it is clear from the above survey that Palladius had been the main stream through which Roman agricultural expertise flowed into the Middle Ages and early Renaissance, and so into the vast *mare nostrum* of modern knowledge.

This translation

This translation is based on the standard edition of the Latin text by Robert Rodgers. My differences from his judgment are listed in the Appendix. Their number may seem large to readers unacquainted with the error-proneness of a manuscript tradition. In fact it is relatively small, given the length of the text and the almost infinite possibilities for copying errors to have occurred in the centuries between Columella and Palladius, and again in the centuries between Palladius and our first surviving manuscripts of his work. The arguments for my emendations are given in Fitch 2013.

In cases where a Latin word has no English equivalent, it sometimes seemed best to coin one, e.g. 'ablaqueate' for *ablaqueare* and 'juger' for *iugerum*. Other less amenable words have been left in their Latin form, e.g. *amurca* and *defritum*. All such words are explained in the Glossary that follows this Introduction, along with terms which are properly English but perhaps unfamiliar to most readers, such as 'shield-budding'.

The numbering system for chapters and paragraphs, found in all modern editions, is not original to Palladius, at least in its entirety, but it has become indispensable for reference. Consequently I have included it in this translation, and where Palladius makes a cross-reference, I have provided the numerical reference in the text or a footnote. It is worth emphasizing that the correct way to cite Palladius' discussion of the 'points' of a stallion, for example, is not by the page-number of this translation, but by book, chapter and paragraph numbers (4.13.2–3), since that system is universally valid for all editions and translations of Palladius.

The lists of contents that preface each Book (not included in this translation) go back to Palladius himself. At a later stage, individual entries from these lists were copied (somewhat inaccurately) into the text, to serve as headings for paragraphs and sections. These headings are sometimes unsatisfactory, and I have therefore used the prerogative of an editor to supplement and adjust them, in order to reflect more clearly Palladius' organization of his work.

My footnotes have the modest goal of elucidating *what* Palladius is saying, not *why* he says it. *What* he is saying is sometimes less clear to a modern audience, even in translation, than it would have been to his contemporaries. To explain *why* he says it, particularly when he disagrees with Columella, would require a detailed commentary making use of both literary and archaeological sources. Such a commentary is a desideratum in Palladian studies, but it is much beyond the scope of this volume.

Glossary

ablaqueate: loosen and pull back the earth around the trunk of a vine or tree, exposing the upper roots (Palladius 2.1).

amurca: olive-lees, the bitter and watery by-product of olive-pressing, useful as a fertilizer, in veterinary medicine, and in other applications.

barley groats (*polenta* or *alphita*): a coarser grind than barley meal (*hordei far*). (Modern 'pearl barley' is somewhat similar, but has been processed to remove the bran layer.)

bud-grafting (*inoculare*): grafting with a single 'eye' or bud. The only method described by Palladius is shield-budding (see below), so the two terms are virtually synonymous in him.

carbunculus: meaning uncertain, apparently a kind of sandstone.

defritum (late spelling of *defrutum*): syrup made by boiling down must. Palladius at 11.18 does not specify a consistency for *defritum*, though he does for *sapa* (below).

first-ploughing (*proscindere*): breaking up the surface layer of the soil into clods. Usually followed by later ploughing or harrowing to produce an even tilth for sowing.

honeywater (*mulsa*): water sweetened with honey.

juger (*iugerum*): very roughly three-quarters of an acre or one-quarter of a hectare (see 'Measures', following).

pith: the *medulla* or relatively soft material at the centre of a plant's stem.

sapa: syrup made by boiling down must to one-third of its volume (11.18).

shield-budding (*inplastrare*, a late spelling of *emplastrare*): a small patch of bark centred on a bud is removed from the host tree, and replaced with a patch ('shield') of the same shape, again centred on a bud, from the donor tree (7.5.2–4).

sextarius: roughly a pint, or half a litre (see 'Measures', following).

treed vineyard: a vineyard where vines are trained up living support trees (usually poplar, elm and ash); described at 3.10.

ulpicum: a large-headed variety of garlic.

Measures

Liquid and dry measures of volume

The basic unit is the *sextarius*, which is approximately 546 ml. It equates roughly to half a litre, or one pint. (A British pint is 568 ml, and an American liquid pint is 473 ml.)

amphora	*8 congii, i.e. 48 sextarii*
modius	*16 sextarii (used only as a dry measure)*
congius	*6 sextarii*
hemina or cotula	*½ sextarius*
cyathus	*¹⁄₁₂ sextarius*

Less common is the *metreta*, a liquid measure equalling about 71 *sextarii* or 39 litres.

Weight

The basic unit is the *libra* (approximately 329 g). Since this does not correspond closely to the modern pound avoirdupois (454 g), I have kept the Roman term. On the other hand, the Roman *uncia* ($\frac{1}{12}$ of a *libra*) is very close to a modern ounce ($\frac{1}{16}$ of a pound avoirdupois), so I have translated *uncia* as 'ounce'.

Length

I have used the English terms 'foot' and 'inch', since the Roman *pes* (296 mm), corresponds closely to the modern foot (305 mm), and both are divided into 12 inches (Latin *unciae*).

cubit	*1½ pedes*
palm	*¼ pes*
inch	*¹⁄₁₂ pes*
finger	*¹⁄₁₆ pes*

Area

The basic unit of area is the juger (*iugerum*). Palladius defines it as 180 x 180 Roman feet (at 2.12, see footnote there), i.e. 32,400 sq. ft.; this is equivalent to 2838 m^2, i.e. very roughly three-quarters of an acre or one-quarter of a hectare.

THE WORK OF FARMING

OPUS AGRICULTURAE

Book 1: Preparations

1. *Preface*

§1 Common sense requires that you first assess the kind of person you intend to advise. If you want to make someone into a farmer, you should not emulate the skills and eloquence of a rhetorician, as most instructors have done. By speaking in a sophisticated way to country folk, they have achieved the result that their instruction cannot be understood even by the most sophisticated. But we must avoid a protracted preface, so as not to imitate those we are criticizing.

§2 We are to speak, if heaven is kind, about every aspect of cultivation, pasturing, farm buildings (following the authorities on construction), and finding water – every detail of activity or nutrition that the farmer needs to observe for the sake of pleasure and production. These topics are distributed throughout the work according to the appropriate season. One exception: on fruit trees I have decided to cover all details of their regimen under the month in which each should be planted.

2. *The four fundamentals*

First, then, the fundamentals of choosing land and cultivating it well consist of four things: air, water, earth, and application. Three of these depend on nature, one on capacity and will. You should examine first the factors belonging to nature: in places you intend to cultivate, the air should be healthy and mild, the water wholesome and easily obtained (whether found on the spot or channelled in or collected from the rain), and the earth fertile and favourably situated.

3 *Assessment of the air*

A wholesome air, then, is indicated by a location well away from valley bottoms and innocent of night mists, and by appraisal of the inhabitants' physique: whether their colour is healthy, their heads steady and sound, their vision unimpaired, their hearing distinct, and whether their voice comes clear and unimpeded from their throat. The benign quality of the air is determined in this way; the opposite symptoms reveal that the atmosphere of the region is harmful.

4. *Assessing the water*

§1 The wholesomeness of the water is recognized as follows. First, it should not be drawn from pools or marshes, it should not originate in mines, it should be translucent and not spoilt by any flavour or odour, and no silt should settle in it. It should also be warm enough to mitigate the frost, but cool enough to temper the fiery heat of summer. But even with all these points observed with regard to appearance, its hidden nature often holds some concealed hazard. So we should ascertain that nature from the health of the inhabitants – if those who drink it have clear throats and healthy heads, and never or rarely suffer complaints in their lungs or chest. §2 For frequently these ailments are transmitted to the lower part of the body by disorders above, so that once the head is affected the cause of the disease descends to the lungs or stomach. (In that case, the air is found to be the more probable culprit.) It is also a good sign if the belly and abdomen and flanks and kidneys are not troubled by any pain or swelling, and if there are no bladder problems. If you see these and similar healthy signs among the majority of the inhabitants, you need have no suspicion of the air or water sources.

5. *Soil quality*

§1 In soils one must look for fertility. There should not be white glebe bare of vegetation, nor lean grit without an admixture of earth, nor pure clay, nor barren sand, nor hungry gravel, nor gold-coloured dust that is thin and stony, nor soil that is salty or bitter or boggy, nor sandy hungry tufa, nor valley land that is too shaded and dense, but a crumbly glebe, almost black, and capable of covering itself with a surface layer of its own grasses – or else it should be of mixed colour. Even if it is loose-knit, it can be bonded by the addition of rich soil. §2 The vegetation on it should not be scabbed or shrivelled or lacking in natural juice. It should bear what is a useful prognostic for grain yield: dwarf elder, rush, reed, grass, clover (not thin), fat blackberries, wild plums.

§3 One should not look particularly for colour, however, but for richness and sweetness. You recognize rich soil thus: you sprinkle a small clod with sweet water and knead it; if it is viscous and sticky, there is clearly richness present there. Again, if you dig a hole and refill it and soil is left over, it is rich; if the soil is insufficient, it is thin; if it tallies and comes level, it is middling. The degree of sweetness, however, is determined if you place soil from a particularly disappointing area of the farm in an earthenware vessel, moisten it with sweet water, and test it by the evidence of its flavour.

§4 In addition, you identify soil that is useful for vineyards by these signs: if its texture is somewhat open and loose, and if the shrubs it produces are smooth, glossy, tall and fertile – wild pears, plums, blackberries and other such – not bent or sterile or drooping in scraggy thinness.

§5 The lie of the land should not be so flat that the soil is stagnant, nor so steep that it washes down, nor so sunken as to lie in the bottom of a deep valley, nor so high as to feel storms and heat excessively. The advantage always lies in a balanced medium between all these situations: either open level country that draws off rainwater by means of a barely perceptible incline, or a hill that falls away gently with sloping sides, or a valley that is only moderately low and wide enough for movement of air, or a mountain protected by another summit, with some shelter from harmful winds, or ground that is high and rough but wooded and grassy.

§6 There are very many types of soils – rich or thin, dense or loose, dry or moist; and most of these are defective, yet often desirable because of the different requirements of seeds. But the first choice, as I said above, is rich loose-textured land, which needs the least labour and returns the greatest yield. Second in value is dense land, which requires very great labour, but does answer one's prayers. The worst kind is dry and simultaneously dense and lean or cold. Such land must be shunned as though it were pestilential.

6. *Application: maxims essential to farming* [1]

§1 After evaluating these factors, which are natural and cannot be amended by human means, you need to deal with the remaining area, that of application. Your best means of attending to this factor will be to keep in mind particularly the following maxims, drawn from all areas of farm-work.

∞ The master's presence promotes the farm's wellbeing.

∞ Soil colour is not a major requirement, as it is an unreliable guide to goodness.

∞ §2 Rely on the best varieties of each woody plant or crop, but only after testing them on your land; you must not put all your hopes on a new variety before testing.

∞ In moist locations seeds degenerate more quickly than in dry ones; repeated selection is required to counter this.

∞ You must definitely have ironworkers, woodworkers, makers of pots and barrels on the farm, lest the farmhands be drawn away from their regular work by the need to visit the city.[2]

∞ In cold places vineyards should be placed facing south, in hot places north, in temperate places east or, if necessary, west.

∞ §3 The calculation of how many workers are required cannot be uniform, since lands are so diverse; familiarity with the soil and region will readily show what numbers should perform any given task, on woody plants or any kind of seedcrop.

∞ Plants in bloom should definitely not be touched.

∞ Seeds for sowing can only be selected well, if a well-selected person undertakes this duty.

∞ In farming, young men are best suited to carrying out tasks, older men to assigning them.

1. Many of these maxims are cast in aphoristic style like old saws, and the organization of this whole section is meandering, as if to distinguish it from systematic exposition.
2. To purchase such items.

∞ §4 In pruning vines three things should be considered: the fruit one hopes for, the growth that will ensue, the placement that will maintain and reinvigorate the vine.[3]

∞ If you prune a vine earlier, you will achieve more growth; if later, a large amount of fruit.

∞ Trees like people should be transferred from bad places to better ones.[4]

∞ Prune more stringently after a good vintage, more leniently after a small one.

∞ In every task involving grafting, pruning or cutting back, use sharp tools of hard iron.

∞ On vines and trees, complete the necessary tasks before the opening of flower and bud.

∞ In the vineyard a digger should remedy what the plough missed.

∞ In places that are hot, dry and sunny one should not trim the vines,[5] since they prefer to be shaded. And when a south-easterly, or some other harmful local wind, scorches the vineyards, we must cover the vines with straw, even if we have to purchase it from elsewhere.

∞ §5 A branch that is luxuriant, green and sterile in the middle of an olive tree should be cut out, as inimical to the whole tree.

∞ Unproductive and pestilential situations must be avoided in equal measure, whether or not these conditions occur together.

∞ In trenched soil nothing at all should be sown between new vines. In the third year the Greeks advise adding any companion crop you like, except cabbages.

∞ The Greek authorities advise that all legumes should be sown on dry soil; only beans should be scattered on moist land.

∞ §6 A person who rents his farm or land to an owner or tenant-farmer holding adjacent land is asking for losses and lawsuits.

∞ The central areas of a farm are at risk, if the outer areas are not cultivated.

3. i.e. the placement of the pruning cuts.
4. i.e. not vice versa, from good places to worse. The word *arbores*, 'trees', can include vines.
5. The reference is to leaf removal in summer (see 6.2.2).

∞ In boggy soil all wheat turns into a variety of soft-wheat after the third sowing.[6]

∞ Three troubles are equally harmful: sterility, disease, neighbours.

∞ §7 Anyone that plants sterile land with vines sabotages both his own work and his outlays.

∞ Level land produces more abundant wine, hills nobler wine; a north wind fecundates vines exposed to it, a south wind ennobles them. So it is a matter of judgment whether to have more or better.

∞ Necessity takes no holidays.

∞ One should sow when the fields have been moistened by rain; nevertheless, if the dry weather is prolonged, seeds once harrowed in will keep more safely in the fields than in the granaries.

∞ The trouble entailed in a journey is equally detrimental to its pleasure and its usefulness.

∞ §8 One who cultivates a farm endures a creditor whose levies are demanding, and to whom he is bound without hope of discharge.

∞ Someone who leaves unbroken soil between the ploughed furrows detracts from his crops and tarnishes the fertility of the land.

∞ A small plot well cultivated is more productive than a large one untended.

∞ §9 You should reject black vines altogether, except in those provinces and of that variety from which *acinaticium* is made.[7]

∞ A longer support promotes the vine's growth.

∞ When a vine is tender and green, do not cut it back with the knife. All cuts in the branches should face away from the bud, so the drops that flow from the cut will not kill it.

∞ The slenderness or stockiness of the vines should determine how many branches the pruner requires them to nourish.

∞ Deep soil, as the Greeks assert, produces big olive trees but berries that are smaller, watery and slow to ripen, more like *amurca*.

6. i.e. when wheat grown on boggy land is used to sow the same land three times in succession.

7. A speciality wine from dried grapes, sweet and concentrated. 'Black' vines are those that produce black (i.e. dark red) grapes and wine.

∞ Olives are helped by a climate that is warm, and open to moderate winds that are not rough or violent.

∞ §10 A vine trained on a bar should be raised over the years so its crown reaches four feet from the ground in more inclement places, but seven feet in more tranquil places.

∞ A garden in a gentle climate with springwater running through it is almost free of limitations, and needs no sowing regimen.

∞ Tying-up should be done when the grapes are unripe, while there is no fear of knocking off the berries or bursting them. §11 The vine-ties should be moved, lest one spot be chafed by the constant presence of the fastenings.

∞ If the digger is seen by the vine's open eye,[8] the great hope of the vintage will be blinded, and so digging should be done while it is closed.

∞ Test the depth of the soil, along with its fertility, to two feet for grain and four for trees or vines.

∞ As a young vine easily puts on growth if lovingly tended, so it meets a swift end if untended.

∞ §12 Assess your abilities and observe moderation when you take on cultivation, lest you outdo your own strength, exceed your limits, and abandon in ignominy what you took on in arrogance.

∞ Seed should not be more than a year old, lest it be spoilt by age and fail to sprout.

∞ Hill crops are admittedly firmer in grain, but they will return less in quantity.

∞ All sowing should be done on a waxing moon and on warm days, for warmth draws things out but cold closes them up.

∞ §13 If you have land covered in useless bush, divide it up so as to turn the rich areas into clean new fields, but leave the sterile areas covered in bush, because the former give a return through their natural productiveness, while the latter are fertilized thanks to fire. But you will diversify the treatment of the areas to be burnt, so as to come back to any burnt land after a five-year interval; in this way you will cause even sterile glebe to compete on equal terms with high yield.

---

8. The eye is a frequent metaphor for the bud, e.g. 4.1.1–2.

∞ §14 The Greeks advise that olives, when being propagated and gathered, should be handled by pure boys and virgin girls; they are mindful, I suppose, that chastity is the presiding spirit of this tree.[9]

∞ It is superfluous to teach the names of grains, which change repeatedly with locale or age. It is sufficient that we select the leading varieties in the area where we farm, or test any that are brought in.

∞ If lupine and fodder vetch are cut green, and one ploughs straightway over their cut roots, they fertilize the fields just like dung; but if they dry out before you plough, the soil's vital moisture is drawn away into them.

∞ Well-watered land needs more dung, dry land less.

∞ §15 In hot, maritime, dry, sunny, level places all vineyard tasks should be started earlier; in cold, inland, wet, shaded, mountainous places, later. I say this not only in regard to months or days, but also the hours of work.

∞ When any rural work is recommended, it is not too early if done 15 days ahead, nor too late if done 15 days afterward.

∞ All crops thrive particularly in open flat ground which is loose and slopes towards the sun. §16 Dense clayey wet ground nourishes emmer and wheat well. Barley delights in loose dry soil, for if it is sown on muddy land, it dies. Three-month crops[10] suit cold snowy places, where the climate in summer is moist; in other places they rarely give a successful return. Three-month seed will give a better return in warm locales, if sown in autumn.

∞ If necessity compels us to have some hope of salty land, it should be planted or sown after autumn, so its badness can be washed out by winter rains. In addition some sweet soil or river sand should be added, if we are entrusting cuttings to it.

∞ §17 We should choose soil of moderate quality in which to establish a seedbed, so the seedlings can be transferred to better soil.

∞ Stones left on the surface are freezing in winter and scorching in summer; therefore they harm crops, bushes and vines.

9. An allusion to the virgin goddess Minerva, patron of the olive tree.
10. These are grain crops adapted to ripen in three months from sowing.

∞ The earth being moved around trees should be changed in sequence, so the lowest level replaces what was on top.

∞ §18 In fertilizing trees we shall make layers, first moving soil near the trunk and then fertilizer, so the potential benefits alternate as the material accumulates.

∞ To supervise the farm you should not appoint any slave for whom you once had a tender fondness, since his reliance on that past love will lead him to expect impunity for any present fault.

The Farm and its Buildings

7. *Choice and site of the farm*

§1 In choosing or buying a farm, you will have to check whether its natural productiveness has been spoilt by the laziness of its cultivators, and whether they have wasted the soil's fertility on inferior plantings. This can be corrected by grafting better stock, but it is better to have unimpaired enjoyment of these things, rather than postponed success and the hope of putting things right. True, there can be an immediate improvement in the seeds of grain crops. §2 But in vineyards one must particularly watch out for and avoid the situation created by many farmers, who think only of prestige and the extent of their trenched ground,[11] but plant vines that are either sterile or of inferior flavour. It will cost you much toil to correct the situation, if you purchase a farm beset by flaws like these.

§3 The actual farmstead to be chosen should be situated as follows. In cold regions the farmstead should be exposed to the east or south: if it is cut off from these aspects by some large mountain standing in the way, it will be freezing, with the sunshine excluded by a northern aspect, or postponed till evening by a westerly one. But in hot regions one should instead choose a northerly aspect, which answers equally well to utility and pleasure and health. §4 If there is a river near a planned construction site, we must check its nature, since a river's

---

11. On deep-digging (*pastinatio*) of land intended for vineyards, see 2.10. This labour-intensive process was expensive, and was therefore seen as a gauge of the farmer's wealth.

exhalations are generally harmful. If such is the case, the builder will have to avoid it. But marshy ground must absolutely be avoided, especially that which faces south or west and dries out in summer, in view of the pestilence or noxious animals that it generates.

8. *The building*

§1 The planned building should be in keeping with the farm's value and the owner's wealth, since it is generally more difficult to sustain ongoing expenditure on an extravagant scale, than it is to build on that scale. The size should therefore be calculated so that if some accident occurs, it can be put right with one year's income, or at most two years', from the farm in which it stands. §2 The site of the main building should be somewhat higher and dryer than everything else, to prevent damage to the foundations and to enjoy a pleasant outlook. The foundations should be laid half a foot wider on each side than the wall to be built on them.[12] If you hit stone or tufa, the business of seating them is easy; it will only be necessary to cut the shape of the foundations into these materials to a depth or one or two feet. If firm solid clay is found, the depth assigned to the foundations should be one-fifth or one-sixth of the intended height above ground. But if the earth is looser, they should be dug to a greater depth, until clean clay appears with no trace of broken stones. If clay is completely absent, it will suit to sink one-quarter in the ground.[13]

In addition, one must ensure that the building can be surrounded by gardens and orchards or meadows. §3 The whole building should be oriented so that its long front faces south, receiving the early morning sun in winter on its corner, while it is somewhat turned away from the setting sun in summer. The result will be that the building is lit by the sun through the winter, and unaffected by its heat in summer.

12. Palladius' recommendations on the width and depth of foundations, not paralleled in surviving ancient sources, are illustrated in Figure 1.
13. Palladius' terseness becomes ambiguous here: he could mean 'one-quarter of the height above ground' or 'one-quarter of the whole wall', i.e. one-third of the height above ground.

9. *Floors for winter and summer quarters*

§1 The design should be such as to provide living quarters for both summer and winter in a small compass. Those intended for winter should be positioned in such a way that they can be gladdened by virtually the whole course of the winter sun. They will need to have suitable paved floors. §2 In constructing these at ground level, the first point to watch is that the joists and boards should be level and firm, lest the vibration caused by unsteady construction upset the steps of those walking on them; second, that oak planks should not be mixed with *aesculus* oak. For if oak gets wet after it has begun to dry, it will warp and make cracks in the floor, whereas *aesculus* is durable and without faults. §3 But if *aesculus* is unavailable and you have a supply of oak, cut the oak thin and place a double layer of boards at right angles to each other, nailed at short intervals. Platforms made of *cerrus* oak or beech or ash will last a very long time, if straw or bracken is spread on them to stop moisture from the lime ever reaching the platform itself.[14]

§4 Then you will lay the basis for the floor on it;[15] you will set down concrete, i.e. two parts broken rock combined with one part lime. After bringing this up to a thickness of six fingers and checking with a rule that it is level, you will need (if these are winter quarters) to lay a floor that the servants can stand on even in bare feet without freezing in winter. So after spreading the concrete or crushed-tile mortar,[16] you will mix charcoal that has been piled up and trodden firm with grit and ash and lime, and have a six-inch thickness laid of this material. Once levelled it will form a black floor and will quickly absorb anything poured on it from cups.

§5 But if they are summer quarters, they should face north and the sunrise at the solstice;[17] they should be given a floor of crushed-

---

14. The oak boards were used to construct a level form on which the concrete floor was poured. *Aesculus* oak is possibly *Quercus frainetto* (*farnetto*), the so-called Hungarian oak; *cerrus* oak is *Quercus cerris*, the Turkey oak.
15. The basis (*statumen*) was a layer of fist-sized rocks, as Palladius' source Faventinus specifies. It looks as if that specification had been lost in Palladius' manuscript of Faventinus. For this layer of rocks cf. 6.11.1 below.
16. The latter is mortar with crushed tile added, i.e. the impermeable *opus signinum*.
17. i.e. north-east.

tile mortar, as mentioned above, or of marble or mosaic or lozenge-shaped stones, whose corners and sides should be flush so as to give a level surface. If you are without these materials, crushed marble should be sifted over the space, or a smooth surface should be made of sand and lime.

10. *Lime and sand*

§1 In addition, it is necessary when building to know what type of lime and sand is useful. Of excavated sand, then, there are three kinds: black, red, white. All are excellent; red is best, white is next in quality, and black takes third place. Of all these, sand that squeaks when gripped in the hand will be useful in construction. Again, if you sprinkle it on a white cloth or piece of linen and shake it off, and it leaves no mark or stain, it is outstanding. §2 But if there is no excavated sand, it will need to be collected from rivers or gravel deposits or from the shore. Marine sand dries more slowly, and therefore building with it must not be done continuously but at intervals, lest the structure fail when a load is placed on it; it also breaks down ceiling plaster with its salty moisture. Excavated sands are useful for plaster and ceilings as a result of their quick drying, and they are better if mixed immediately after being dug out; for they become insubstantial if exposed to lengthy sunshine or frost or rain. River sands will be found more suitable for plaster. §3 But if marine sand must be used, it will be beneficial to drop it first in a freshwater pool, to get rid of the problem of the salt by rinsing it in sweet water.

As for lime, we shall fire it from hard white rock or travertine or river marl or red rock or pumice or lastly marble. That made from dense hard rock is appropriate for structural work, whereas that from softer or porous stone is more usefully applied in plaster. One part lime should be mixed with two parts sand. In the case of river sand, however, if you add one-third of sifted shards, the result will be a remarkable solidity in the work.

11. *Protecting brick walls*

If you want to use brick walls for the main building, you will need to ensure that, after the walls are finished, a tile structure is added with projecting cornices, a foot and a half in height, on the top under the roofbeams. In this way, if the roof-tiles or capping-tiles[18] are damaged, the leaking rainwater cannot penetrate the wall.[19] Then you must see to it that the brick walls are dry and roughened when the finishing coat is applied to them. It cannot stick to walls that are moist or smooth, and so you will have to give them three preliminary coats, so they can take the finishing coat without it being spoilt.[20]

12. *Light and height*

In rural construction we must ensure first and foremost that it is bright and well lit; then that we orient the seasonal quarters, as I said above (1.9), in the appropriate directions, i.e. summer quarters north, winter quarters south, spring and autumn quarters east. The dimensions to be observed in dining-rooms and other rooms are these: the length and breadth should be added together, and half that total should be taken as the height.

13. *Cane-based ceilings*[21]

§1 In rural buildings it will be useful to fashion ceilings from material that is easily found on the farm. Accordingly we shall make them either of boards or of canes, as follows. We shall set joists of Gallic wood[22] or cypress horizontally along the main axis of the space where the ceiling is to be made, with a gap of a foot and a half between them. Then we shall fasten them to the frame above them[23] with wooden

18. Semi-cylindrical tiles placed over the joints between the roof-tiles.
19. Bricks were either sun-dried, as Palladius describes 6.12, or fired at relatively low temperatures. Consequently a coping of impermeable tile was needed to prevent water entering the top or face of a brick wall and destabilizing it. See Figure 2 for illustration.
20. The multiple coats are described in more detail in 1.15 below.
21. This paragraph is illustrated in Figure 3.
22. Fir – possibly a superior type of fir (12.15.1, Pliny 16.197), to meet Vitruvius' criticism that fir decays too quickly (7.3.1).
23. i.e. the transverse beams supporting whatever is above.

ties of juniper or olive or box or cypress, and run two rods along each space between the joists, fastened with cords. §2 Next we bind up marsh canes, or that thicker type that is commonly used, after flattening them, forming a lattice and tying it tightly, and fasten them over the entire area to those joists and rods. Then we shall first coat all this with a layer of pumice mortar and smooth it with a trowel, to fasten the individual canes together. Next we shall smooth it with sand and lime. Thirdly we shall spread a mortar of crushed marble powder mixed with lime, and buff it to a high sheen.

14. *Testing lime for stucco*

People are often keen on stucco work too. We shall have to use lime for it, after slaking for a long time. To test its usability, you will chop the lime with an axe like wood. If the edge of the blade does not strike against anything, and if what adheres to the axe is soft and sticky, it is clearly suitable for stucco work.

15. *Wall cladding*

Here is how to make the cladding of the walls strong and shiny. The first coat should be worked frequently with trowels. When it has begun to dry, a second coat should be applied, and then a third. After these three layers, a coat of crushed marble plaster should be applied with the trowel. This plaster should have previously been worked for long enough that the shovel used to work the lime is clean when we lift it out. When this crushed marble coating also begins to dry, another thinner layer should be applied. In this way it will keep both its solidity and its sheen.

Cisterns, and caulk for cold water

16. Now one must avoid what many have done in order to have a water supply, namely burying their farms at the bottom of valleys, and placing a few days' pleasure ahead of the health of the inhabitants. We shall fight shy of doing this all the more, if the region where we live is prone to diseases in summer. If there is no spring or well, it will be advantageous to construct cisterns, in which water can be

collected from the roofs of all the buildings. They are made in the following way.

17. §1 The walls should be of waterproof mortar,[24] the size to be built should suit your wishes and resources, and the length should exceed the width. Its base should be solidified with a deep layer of concrete and smoothed by pouring a floor of crushed-tile mortar, leaving space for drains. This floor must be polished assiduously to a sheen, and rubbed repeatedly with rich animal fat that has been boiled down. §2 When the moisture has dissipated and the floor has dried out, so that cracks will not open anywhere, the walls too should be covered and coated with a similar layer; and so, after the work has dried and solidified for a long time, the water can be accommodated. Naturally it will be advantageous to place eels and river fish here and feed them, so that as they swim the standing water can imitate the movement of a running stream.

But if at some time the floor or wall surface should give way at any spot, we shall apply the following kind of caulk, to keep in the water that is trying to escape. §3 (We shall repair cracks and holes in cisterns, fishponds and wells in this way, even if the water is seeping through the rocks.) You take as much liquid pitch as you wish, and the same amount of the grease that we call axle-grease or tallow. Next you will mix the two in a pot and heat till bubbling, then remove from the fire. When this mixture has cooled, you will add lime gradually and make the whole blend uniform. §4 And when you have produced something like strigil-scrapings,[25] you insert it in the places that are damaged and leaking, and press and ram it in very firmly.[26]

It will be healthful to conduct water here through earthenware pipes and guide it into the covered cisterns. For rainwater is preferred to all other kinds for drinking, so that even if flowing water can be brought in, it should be reserved, if it is unwholesome, for the baths and for garden cultivation.

24. *Opus signinum*, a mortar made waterproof (and pink) by the addition of crushed terracotta.
25. The mixture of oil, sweat, skin and dirt scraped off by the strigil; the medical writer Celsus likewise uses strigil-scrapings to specify the consistency of mixtures.
26. Other recipes for caulk are given in 1.40 below.

18. *The wine-room*[27]

§1 We should have the wine-room facing north; it should be cold and almost dark, a long way from baths, animal stalls, the furnace, dung-heaps, cisterns, water and other foul-smelling places. It should be equipped with necessities so as not to be overwhelmed at vintage-time, and arranged in such a way that (basilica-like) it has a pressing-floor built at a higher level,[28] to which one can ascend by three or four steps between two vats which are sunk on each side to receive the wine. From these vats there should be built-in channels or earthenware pipes running round the sides of the walls, with pots alongside them at a lower level into which the flowing wine can pour through abutting pipes.[29] §2 If the quantity is too large, the middle of the cellar will be assigned to casks. Lest they block the walkways, we can place these casks on higher 'donkeys'[30] or above the pots, with a generous space between them, so that access is available, if needed, for the person in charge. If we assign a particular area to the casks, this area should be made impermeable like the pressing-floor with low surrounding walls and a floor of crushed-tile mortar, so that even if a cask breaks without being noticed, the escaping wine will be collected in the reservoir beneath, rather than being lost.

19. *The granary*

§1 The site of the granaries requires the same exposure, but it needs in addition to be fairly high, well away from all moisture, manure and animal stalls – cold, windy and dry. The structure needs to be planned with care, so it cannot fracture and crack. Accordingly all the ground should be paved with two-foot or smaller bricks,[31] which we should set on an underlying floor of crushed-tile mortar. §2 Then we shall make separate compartments, if a large quantity of seed is expected, and earmark them for the individual types of grain. If the

27. For an illustration of Palladius' recommendations here, see Figure 4.
28. The point of comparison is that a basilica had a raised dais (*tribunal*), rectangular or apsidal, at one end.
29. See Figure 4. Palladius is our only written source for this neat gravity-fed system.
30. A figurative term for a kind of stand.
31. Two-foot bricks are two foot square.

land is poor and promises smaller yields, the granaries should be sectioned off by hurdles; if the produce is meagre, we shall pile it in wickerwork containers.

After the granaries are built, the walls are coated with a mixture of *amurca* and mud, to which dry leaves of olive or wild olive are added in place of chaff. Once this coat has dried, it is sprinkled again with *amurca*. After this has dried, the grain can be stored. This procedure deters weevils and other vermin. Some people mix leaves of oleander with the grain to help in preserving it.

But nothing will contribute more to preserving the grain for a long time than to pour it from the threshing-floors into another nearby place and cool it for some days, and then bring it into the granaries. §3 Columella says that the grain should not be tossed to air it, since this contributes to mixing insects throughout the heaps.[32] If they are not disturbed, they will stay within a palm's breadth of the surface; while this veneer, so to speak, is spoilt, the remainder will keep unharmed. The same writer states that noxious animals cannot breed below the aforementioned depth. Dried stinkwort[33] spread beneath the grain prolongs its life as the Greeks assert. The granaries should face away from southerly winds.

20. *The oil plant*

The oil room should face south, and it needs to be protected from the cold. So there must be windows of translucent stone[34] to let in the light. In this way the winter tasks will not be hampered by chilly weather, and, thanks to the warmth, when the oil is pressed there will be no chance of it congealing from cold. For the olive-mills and wheels and press, the shape dictated by custom is well known. The receptacles for the oil must always be clean, lest they spoil the fresh flavour if impregnated with ancient mustiness. If a person is keen to take extra measures, he can build an elevated floor with criss-cross

32. Columella 1.6.16–17.
33. *Conyza* in Greek and Latin: formerly classified as *Inula viscosa* or *graveolens*, currently as *Dittrichia v.* or *gr.*
34. Such translucent panes (*specularia*), made chiefly of mica and gypsum, were also used in greenhouses, etc.

'rabbits'[35] beneath it, and provide heat by lighting the furnace. In this way the oil room will be warmed by clean heat without the smell of smoke. Oil is often permeated by smoke, and spoilt in colour and flavour.

21. *Stables for horses and oxen*

Stables for horses or oxen should have a southerly aspect, but they should not be without sources of light on the north side; when closed in winter these do no harm, and when opened in summer they provide cool. For the sake of the animals' hooves, the stables should be raised above any moisture. Oxen will become sleeker if they have a fireplace close by and face its light. Eight feet width is plenty for the standing of each pair of oxen, and fifteen feet in length. Oak boards should be laid on the floor of the horses' stalls, to make it soft enough for lying and hard enough for standing.

22. *The farmyard*

The farmyard should be open to the south and exposed to the sun, since it will be quite easy to fashion lean-tos, using forked sticks, horizontal poles and leaves, to moderate summer's heat for the animals in the yard. The lean-tos will be covered with shingles, or with tiles if a supply is available, or (if an easier and free material is preferred) with sedge and broom.

Aviaries and Raising Fowl

23. *Bird houses*

Bird houses should be constructed along the sides of the farmyard walls, since bird dung is especially important in agriculture, with the exception of geese droppings which are quite harmful to everything. But homes for the other birds are very important.

---

35. These are the underfloor tunnels of a hypocaust system.

24. *The pigeon cote*

§1 Support for the pigeon cote can be provided by a high little tower erected in the farm headquarters with smooth whitened walls. There should be openings facing all four directions, as is customary, of very small size, so as to allow entrance and egress only to the pigeons. §2 Nesting-boxes should be fashioned inside. They will be safe from weasels if old esparto-rope, such as is used for shoeing animals, is scattered among them, as long as someone brings it secretly without others seeing. They do not die off or abandon the house if you hang up a piece of the strap or belt or rope from a strangled man in each of the openings. They bring in other birds if they are regularly fed cumin or if their wing-pits are touched with unguent of balsam. §3 They lay frequently if they are often fed roasted barley or beans or bitter vetch. For thirty flying pigeons[36] three *sextarii* per day of wheat or pollard will suffice, as long as we provide bitter vetch in the winter months to encourage laying. One should hang up twigs of rue in many places against harmful animals.

25. *Turtle-doves*

Below the chamber of the pigeon cote two cubicles may be made; one, in which turtle-doves can be enclosed, should be small and almost dark. It is very easy to feed them, for they require nothing except a steady supply in summer (the only season when they grow very fat) of wheat or millet steeped in honeywater. One half-*modius* per day is enough for 120 turtle-doves. Of course clean water should be provided for them frequently.

26. *Thrushes*

§1 The other cubicle is to be used for feeding thrushes. If they are fattened out of season,[37] they yield both the pleasure of their food and a very good return: frugality benefits from extravagance![38] One

36. 'Flying' probably designates birds that can come and go to the fields to find food, as distinct from birds in suburban situations that need to be confined.
37. i.e. in summer. The fieldfare, *Turdus pilaris*, wintered in Italy in large numbers, and was therefore inexpensive at that season, but scarce and expensive in summer.
38. Countryfolk would not go to the extravagance of raising thrushes out of season

makes a place that is clean, bright and smooth on all sides. Here rods are fastened horizontally for the birds to perch on after flying in their enclosure. In addition green branches should frequently be changed. §2 Figs mashed and mixed with flour should be offered very generously. Also berries of myrtle, if you have a supply, and of mastic, wild olive, ivy and arbutus should sometimes be offered to prevent boredom with their feed, and most of all clean water. They should be put in the enclosure, undamaged and shortly after capture, together with a certain number that have been fed for some time: the company of these birds will allay their fear and dejection at their new captivity, so they will take food.

27. *Chickens*

§1 Any woman whose nature is at all industrious knows how to raise chickens. Let it suffice to give this advice about them, that they should have access to smoke,[39] dust and ashes. Preferably they should be black or yellow in colour: whites should be avoided. They become sterile if fed the marc from wine-pressing. Half-cooked barley will induce them to lay often, and produce bigger eggs. Two *cyathi* of barley is ample food for one free-range hen.[40] Eggs should always be placed under them in uneven numbers, and on a waxing moon – from the tenth to the fifteenth day of the moon.

§2 Pip sometimes occurs among them: it coats the end of the tongue with a white skin. This is lightly picked off with the fingernails and the spot touched with embers, and the cleaned wound is sprinkled with crushed garlic. Again, a morsel of garlic crushed with oil is placed in their throat. Also stavesacre is helpful if regularly mixed with their food. If they eat bitter lupine, its seeds emerge under their eyes. Unless these are removed with a needle by lightly opening the skin, they kill them. §3 We may tend the eyes with purslane juice or woman's milk applied externally, or ammoniac salt with which honey and cumin are mixed in equal quantities. Their lice are killed

---

for themselves, but they benefit from the demand for such luxury items on the part of the wealthy.

39. 'The hen-house should adjoin the furnace or kitchen so that the smoke will reach the birds, since it is particularly beneficial to them' (Columella 8.3.1).

40. As distinct from a bird enclosed in the henhouse.

by stavesacre and parched cumin in matching amounts, ground up together with wine and bitter-lupine water,[41] as long as it penetrates deep into their feathers.

28. *Peafowl*

§1 It is very easy to rear peafowl, unless you face a threat to them from thieves or predatory animals. For the most part they range through the fields and feed and raise their chicks unaided; at dusk they make for the highest trees. But they require your attention in one regard, namely to protect the females from the fox when they are brooding, which they do anywhere in the fields. For that reason they are raised with better success on small islands. Five females are enough for one male. §2 The males attack the eggs and their own chicks as if they were unrelated, until the chicks grow their distinctive crest. They become lusty after mid-February. Lightly roasted beans stimulate their sex instinct, if fed to them warm every fifth day: six *cyathi* are enough for one bird. The male reveals his desire to mate when he curves his jewelled tail like a cloak around him, and displays the eyed top of each feather in its own place, while rushing forward with a screech.

§3 If the peafowl eggs are placed under hens, the mothers, exempted from incubation, lay three times per year. The first laying is usually of five eggs, the second of four, the third of three or two. If this is your plan, nurse-hens should be chosen: for nine days from the moon's first increase, they will have nine eggs to sit on, five from peafowl and the rest from their own kind. §4 On the tenth day all the hens' eggs should be removed, and replaced by the same number of fresh hens' eggs, so that these can hatch with the peafowl eggs on the thirtieth of the moon (i.e. after 30 days have passed). Peafowl eggs placed under a hen should frequently be turned by hand, since she herself will hardly be strong enough to do this. You will mark one part of the egg, so as to know you have turned it at intervals. You should choose larger hens, for you can place fewer eggs under smaller ones.

§5 Once they are hatched, if you wish to transfer them from several hens to one, Columella says 25 is the right number for one

41. Water in which that herb has been boiled.

nurse;[42] but I think that in order for them to be reared well, 15 is enough. For the first days barley meal sprinkled with wine will be given to the chicks, or a mash of whatever grains, cooked and cooled. Later chopped-up leeks will be added, or fresh cheese (but pressed, since whey is harmful to chicks). Locusts too are provided, with feet removed.[43] §6 They should be fed this way till the sixth month. Then barley can be provided in the customary way. Also 35 days after hatching they can safely be let out into the field in the company of their feeding-nurse, whose clucking summons them back to the farm. You will fend off pip and digestive problems with the same remedies used to cure chickens. The greatest danger for them is when their crest begins to grow, for they languish rather like infants when their baby teeth push open the swollen gums.

29. *Pheasants*

§1 In raising pheasants one must make sure to obtain new birds for propagation, i.e. those born the previous year, for old birds cannot be productive. The females are bred in March or April. One male serves for two females, since he does not equal other birds in libido. They will produce offspring once a year. The sequence of laying ends after about 20 eggs. §2 It will be better for hens to incubate them, with one nurse-hen covering 15 pheasant eggs; the others placed under her should be of her own kind. In this process we should observe what was said about other birds with regard to the moon and dates (1.28.3–4). The thirtieth day will bring the fully-developed chicks into the light. For 15 days they will be fed barley meal lightly cooked and cooled, sprinkled with a drizzle of wine. Later you will provide cracked wheat and locusts and ant eggs. They should definitely be kept from access to water, to prevent the pip from finishing them off. §3 If pheasants do suffer from pip, you will have to rub their beaks frequently with garlic ground up with liquid pitch, or else remove the infection as is regularly done with chickens.

§4 The system of fattening is as follows. Meal from one *modius* of wheat, kneaded into very small pellets, is the right amount to provide

42. Columella 8.11.13.
43. i.e. the lower legs, which are barbed.

to a penned pheasant over 30 days; or if you want to provide barley meal, the meal of one and a half *modii* over the prescribed period will complete the fattening. One must of course make sure that these pellets are lubricated with a sprinkling of oil before insertion in their throats, lest they drop under the base of the tongue. If that happens, they will perish straightway. We should also take great care that rations are digested before fresh are provided, since a load of undigested food very easily kills them.

30. *Geese*

§1 It goes without saying that the goose is not easily sustained without water, nor without grass. It is inimical to seeded areas, because its biting and its dung damage the crops. It provides goslings and feathers, which we can pluck both in autumn and in spring. Three females serve for one male. If there is no river, a pond must be formed; if grass is not available, we shall sow clover, fenugreek, wild chicory and lettuce as feed. Whites are more fertile, parti-coloured or dark ones less so, because they have crossed over from the wild type to the domestic. §2 They sit from the first of March up to the summer solstice. They will lay more, if you put the eggs under hens. We leave the last clutch for the mothers to rear since they are about to have a rest. (When about to lay they should be led to the pen; after you have done this once, they will maintain the habit by themselves.) You will set goose eggs under hens just as you do peafowl eggs (1.28.3–4). But in order to prevent the goose eggs being damaged, nettles should be laid beneath them when they are set under the hens. §3 The young should be fed inside for their first 10 days; then we can take them out in calm weather, somewhere where there are no nettles, since they are vulnerable to their sting.

They are best fattened at four months, for they put on weight better at a young age. Barley groats will be given thrice daily. Freedom to wander far and wide is precluded. They will be enclosed in a dark, warm place. In this way even the bigger ones fatten in the second month; the little ones often do so by the thirtieth day. They fatten better if we provide as much soaked millet as they can eat. Among foods for geese all kinds of legumes can be offered except bitter vetch. §4 We must also ensure that the goslings do not swallow hairs from kids. The Greeks in fattening geese blend two parts barley groats and four of bran with warm water, and heap it up to be taken at will by the feeding bird. They supplement with drink thrice a day. They also provide water in the middle of the night. After 30 days have passed, if you wish to soften their liver, you will mash figs, soak them in water and roll them into little nuggets, and supply them to the geese for 20 days on end.

Farm Facilities and Equipment

31. *Ponds*

Once these matters have been organized, there are others to deal with. There will need to be two ponds near the farmhouse, either sunk in the ground or of cut stone. It is easy to fill them from a spring or rainwater. One of them is to be used by livestock or waterbirds, while the other is for soaking twigs and hides and lupines and everything else that requires moistening in the routine of country life.

32. *Storage*

It does not matter where the repositories for hay, straw, wood and canes are placed, so long as they are dry, well-ventilated and a long way from the farmhouse, because of the danger of fire starting unnoticed in those materials.

33. *The dungheap*

§1 The stockpile of dung will need to have its own place, one that has plentiful moisture and faces away from the main building because of the awful smell. The plentiful moisture will ensure that any thornbush seeds in the dung will rot. Donkeys' dung is of first quality, especially for gardens; then that of sheep and goats and beasts of burden; pigs' is worst. Ashes are excellent. The dung of pigeons is hottest, and that of other birds is quite useful, except water birds. §2 Dung that has rested a year is useful for grain fields, and does not produce weeds; if older, it will be less useful. For meadows, however, fresh dung will promote the richness of the grass. Waste from the sea,[44] if rinsed in fresh water, will act as a substitute for dung when mixed with other material; likewise the mud spewed out by gushing water or a flooding river.

34. *The garden: location, fencing and sowing*

§1 The gardens and orchards will need to be close to the house. The garden should ideally be at the foot of the dungheap, so as to be fertilized automatically by the liquid from it; but sited well away from the threshing-floor, since the dust from the chaff is injurious. A favourable location is on flat ground with a gentle slope that will lead running water through the various zones of the garden. §2 If there is no spring, you must either sink a well, or, if you cannot do so, build a pool at a higher level, so that its supply of water from rainfall can irrigate the garden during summer's heat. If you have none of these resources, you will constantly dig over your garden quite deeply, to three or four feet, as in trenching;[45] with such cultivation it can disregard droughts. §3 Although any kind of soil is suitable for the garden, if supplemented as needed with dung, nevertheless these kinds should be avoided in making your choice: the clay that we call argil, and red clay. In gardens that receive no help from natural moisture, you will also take care to create separate sections, a garden for winter cultivation facing south and a summer garden facing north.

§4 There are many forms of protection. Some people, by enclosing mud between forms, make walls comparable to those laid piece by piece.[46] Those who have the materials raise garden walls of mud and stones. Most pile up rocks without mud and arrange them in lines. Some surround the areas to be cultivated with a ditch; this should be avoided (since it draws moisture away from the garden) unless the cultivated area happens to be marshy. Others set out young thorn plants and seeds as protection. §5 But it will be better to collect ripe seeds of bramble and the thorn called dog-bramble,[47] and mix them

44. i.e. seaweed washed onto the beach.
45. The comparison is with *pastinum*, land trenched in preparation for planting a vineyard (see 2.10).
46. i.e. of separate bricks.
47. Thought to be our dog rose, *Rosa canina*, which has sharp hooked prickles.

with flour of bitter vetch moistened with water; then smear old *spartum*[48] ropes with this mixture, so the seeds, lodged in the ropes, will keep until the beginning of spring. §6 Then we shall make two furrows where the hedge is going to be, three feet apart and 18" deep, lay the ropes with their seeds along them and cover them lightly with earth. So on the thirtieth day the briars come up, and you need to help them while tender with supports. They will merge together across the space that was left empty.[49]

§7 There should of course be separate areas in the garden, so that those to be seeded in autumn can be trenched in springtime; those that we intend to fill with seed in spring, we should dig over in autumn. Thus each trenched area will be 'cooked' by the beneficial effects of cold or sun.

The seedbeds should be made long and rather narrow, i.e. twelve feet in length and six in width, sectioned off in this way to permit weeding from each side. Their sides should be two feet high in wet or well-irrigated places; in dry places a height of one foot will suffice. But if the moisture tends to flow away between the beds,[50] the spaces between them will need to be higher than the beds themselves, so that moisture let in from above can enter more easily, and, after soaking the thirsty bed, can be shut out and diverted to other beds.

§8 Though we indicate specific times for sowing month by month, each individual should observe them with reference to the nature of the place and climate. In cold places the autumn sowing should be brought forward, and the spring sowing delayed; but in hot regions the autumn sowing can be done later, and the spring sowing earlier. All sowing should be done on a waxing moon, but all reaping or gathering on a waning moon.

48. This name is applied to two plants whose fibrous stems were used for making rope: Spanish broom and esparto-grass.
49. i.e. the gap of three feet between the two furrows.
50. i.e. if the irrigation water runs along the walkways rather than soaking into the beds.

35. *Remedies for use in the garden or the farm*

§1 *Against mists and the rust* you will burn piles of straw and rubbish set out at many spots around the garden, all at once, when you see a threat of mist.

Against hail many things are prescribed. A millstone is covered with a brown-red cloth. Again, bloody axes are raised threateningly against the sky. Again, the whole area of the garden is encircled with white bryony, or an owl is fastened up with its wings stretched open, or iron tools are smeared with bear grease. §2 Some keep a supply of bear fat pounded with olive oil, and grease their billhooks with it when about to do the pruning. But this remedy must be kept secret, so no pruner understands it. Its power is said to be such that it cannot be harmed by frost or mist or any animal. It is important to note that if divulged the procedure has no force.

Against fleas[51] *and slugs* we spread either fresh *amurca* or soot from ceilings.

Against ants, if they have an entrance-hole in the garden, we should place the heart of an owl close by; if they come from outside, we shall outline the whole garden area with ash or white chalk powder.

§3 *Against caterpillars* the seeds to be sown should be steeped in houseleek juice or caterpillars' blood. Chickpea should sown among the vegetables for its great efficacy. Some sprinkle fig tree ash over the caterpillars. Again they sow squill in the garden, or at the very least hang it up. Some have a menstruating woman walk around the garden wearing no fastenings anywhere, her hair loose and feet bare, against caterpillars and other pests. Some fasten river crabs on crosses at several points within the garden.

§4 *Against the creatures that harm vines*, you will drop the caterpillars that we commonly find on roses into olive oil and let them decay and dissolve, and when the vines are to be pruned, you will smear the billhooks with this oil.

Bedbugs are killed by smearing the beds or rooms with *amurca* and ox bile, or with ivy leaves pounded in oil, or by burning leeches.

---

51. These are the same as the vegetable fleas of §5: presumably our flea beetles (subfamily *Galerucinae*), so called because they jump. *Pulices* 'fleas' is my emendation for the *culices* 'gnats' of the manuscripts.

§5 *To prevent your vegetables from engendering harmful creatures*, dry all the seeds you are going to sow in a tortoise shell, or sow mint in several places, especially among the cabbages. People say bitter vetch has this effect if you scatter a certain amount of it, particularly where radishes and turnips are growing. Alternatively sharp vinegar mixed with henbane sap is said to kill vegetable fleas, if you sprinkle it.

§6 *Caterpillars* are said to be vanquished by burning garlic stalks without the heads throughout the garden area, and generating a strong smell in many places. If we are concerned for our vines, they say the pruning hooks should be smeared with ground-up garlic. They are also prevented from multiplying if you burn bitumen or sulphur around the trunks of the trees or vines, or if you boil caterpillars from a neighbouring garden in water and pour this over all areas of your own garden.

To prevent *blister-beetles* from harming vines, they should be ground up on the whetstone used to sharpen the billhooks.

§7 Democritus [52] states that no harm can come to trees or crops of any kind from any creatures whatsoever, if you place plenty (not less than 10) of river crabs – or sea crabs, which the Greeks call *paguri* – in an earthenware vase with water, cover it, set it in the open to be heated by the sun for 10 days, then drench whatever you want to be unharmed, and repeat this at eight-day intervals until the plants you hoped for are thoroughly established.

§8 *Ants* you will drive away by sprinkling their entrance-hole with ground-up oregano and sulphur. This is harmful to bees too. Ditto if you burn empty snail shells and press the ash into their entrance-hole.

Gnats are driven off by an infusion of galbanum or sulphur; *fleas* by spraying *amurca*, or wild cumin ground up with water, frequently over the floor, or by repeatedly pouring on wild cucumber [53] seed broken up in water, or else water from boiled lupines, fortified with the bitter sap of bryony.

§9 *Mice*, if you pour thick *amurca* in a shallow dish and set it down in the house at night, will stick to it. Again, they will be killed

52. Palladius' attribution of this remedy to Democritus comes from *Geoponika* 5.50 and 10.89.1.
53. *Ecballium elaterium*, squirting cucumber.

if you mix black hellebore into cheese or bread or fat or barley groats, and put it out for them. And a dish of wild cucumber and colocynth will do the same damage.

Against field mice Apuleius[54] states that seeds should be steeped in ox gall, before you sow them. Some people block their holes with oleander leaves; they perish through nibbling these while struggling to get out.

§10 *Moles* are tackled as follows by the Greeks. They instruct that a hole should be bored in a nut, or in some kind of fruit of comparable solidity; an adequate quantity of chaff and cedar-resin should be packed into it, along with sulphur. Then all the moles' little entrances and other air-passages should be carefully blocked; one hole, a wide one, should be left, and at its entrance the nut, with a fire lit inside, should be placed so as to receive a flow of air from one side and emit it on the other side. With their burrows filled with smoke in this way, the moles either take flight straightway, or are killed.

§11 *Country mice*:[55] if you douse their entrance-holes with oak ash, from the frequent contact the scab will attack and kill them.

Snakes are driven off by almost anything bitter, and the disinfection of harsh-smelling smoke repels their infected breath.[56] We should burn galbanum or stag's horns, lily roots, goat's hooves. In this way noxious pests are kept away.

§12 The view of the Greeks is that if a cloud of *locusts* suddenly arises, it can pass over as long as all humans stay hidden inside their buildings. If they catch humans unexpectedly in the open air, no crop is harmed if everyone rushes indoors immediately. They are also said to be driven off by water boiled down with bitter lupine or wild cucumber, if mixed with brine and poured on them. Some people reckon that locusts or scorpions can be driven away if some of them are burnt in the open.

§13 *Caterpillars*: some drive them off with fig tree ash. If they persist, boil together ox urine and *amurca* in equal quantities, and sprinkle all the vegetables with a drizzle of this once it has cooled.

---

54. The attribution is from *Geoponika* 13.5.1.
55. The distinction (if any) between these and the field mice of §9 is unclear. The remedy mentioned here is directed against house mice in *Geoponika* 13.4.2.
56. Their breath was thought toxic in itself, e.g. Columella 8.5.18.

Prasocorides is the Greeks' name for creatures that frequently damage gardens.[57] For them you should take the belly of a wether just killed (still full of its cud) and cover it lightly with soil, in an area where they abound. After two days you will find these creatures amassed there. After doing this two or three times, you wipe out the whole tribe that was doing damage.

§14 *Hail* is believed to be checked if a person, when he sees the danger is imminent, carries the skin of a crocodile or hyena or seal around his property and hangs it in the entrance to the farmhouse or farmyard. Again, if a person carries a marsh turtle on its back in his right hand and walks around his vines, then returns in the same way and sets it on the ground, pressing clods of earth against the curve of the shell so that it cannot turn over but remains on its back, it is said that a menacing cloud hurries past an area protected by this action. §15 Some, when they see the danger threatening, hold out a mirror to catch a reflection of the cloud, and avert the cloud by this remedy: either it is upset at being set against itself, or it makes way for the other cloud as its double. Again, if a sealskin is thrown over one vine in the middle of the vineyard, it is believed to screen the limbs of the whole vineyard against the imminent danger.

§16 All seeds for the garden or fields are said to be kept safe from *all dangers and pests*, if they are first steeped in the mashed-up roots of wild cucumber. Again, the skull of a mare (but one that has bred) should be placed in the garden – better still, a she-ass. Its presence is believed to bestow fertility on all that it faces.

36. *The threshing-floor*

§1 The threshing-floor should not be far from the farmhouse, both for convenience in carrying the grain away and to lessen the danger of theft, since the master or manager may be close by. It should be paved with flagstone or cut in the rock of the hillside, or else packed hard just before threshing time by beasts' hooves with the addition of some water, then shut up and protected with stout hurdles against the

57. The Greek name means 'leek-cutter': perhaps the leekbane or mole cricket, currently in the genus *Gryllotalpa*.

animals which we bring in when threshing.[58] §2 Nearby there should be another level clean space, onto which the grain can be poured so as to cool before being carried to the granaries. This procedure will improve its durability. Then, especially in wet regions, there should be a roof nearby all round, under which the grain can quickly be placed, whether cleaned or half-threshed, if the need arises because of sudden showers. The floor should be in an elevated spot open to the winds on all sides, but well away from gardens, vineyards and orchards. For while dung and chaff benefit the roots of plants, if they settle on the leaves they make holes in them and cause them to dry up.

37. The bees' encampment

§1 We should position the bees' quarters not far from the master's house, perhaps in a portion of the garden that is secluded, sunny, sheltered from winds and quite hot. The hives should be set up in a squared area,[59] to repel thieves or the approach of humans or beasts.

The area should abound in flowers, which need to be diligently provided in the form of herbs or shrubs or trees. §2 Herbs that it should support include oregano, thyme, creeping thyme, savory, lemon balm, wild violets, asphodel, *citreago*,[60] marjoram, the 'hyacinth' that is called iris or gladiolus because of the similarity of its leaves, saffron and all other herbs of the sweetest scent and flower. Among the shrubs there should be roses, lilies, yellow violets, rosemary, rock-rose. Among the trees, jujube, almond, peach, pear, and the fruit-bearing trees from which no bitterness results when the flower is sucked. Wildlings include acorn-bearing oaks, terebinth, mastic, cedar, lime, the smaller gorse, and laurustinus; but yews should be kept away as detrimental. §3 The nectar of thyme yields honey of the best flavour; *thymbra*,[61] creeping thyme or oregano, that

---

58. Horses or bovines may be used to solidify the threshing-floor, but should then be kept off it until threshing-time, when they can be used to thresh the grain by treading it with their hooves. On preparation of the threshing-floor see also 7.1.
59. i.e. a square enclosure (cf. §7).
60. The name suggests a plant smelling like citron: often identified as lemon balm, *Melissa officinalis*, but that seems to be covered earlier in Palladius' list by *mellisfillum*.
61. Probably a particular species or variety of savory.

of second quality; rosemary and savory, that of third quality. Other plants like arbutus and vegetables produce the flavour of country honey. The trees should be set out on the north side; shrubs and bushes should line up in rows under the garden walls; then we shall sow the herbs on flat ground beyond the shrubs. A spring or stream should be fitted in here – a slow-moving one that will form shallow pools as it passes through; these should be covered with some open brush laid horizontally, to provide safe places for the bees to alight when thirsty.

§4 The bees' encampment should be a long way from everything that produces foul smells: bath-houses, animal stalls, kitchens, drains. We must also keep away creatures that are detrimental to bees: lizards and beetles and suchlike. In addition we should frighten off birds with rags and rattles. Their keeper should pay them frequent visits in a clean and chaste state, and should have new hives prepared in advance to catch the inexperienced young of the swarms. §5 The smell of muck should be avoided, burnt crabs,[62] and any place that responds to the human voice with a false imitation. These herbs too should be absent: spurge, hellebore, thapsia, wormwood, wild cucumber, and all bitterness that is adverse to the creation of sweetness.

§6 The best hives are those formed of cork pulled from the cork-oak, since they do not transmit fierce cold or heat. They can be made of fennel stalks too. If these materials are unavailable, they should be fashioned from willow withies or the wood of a hollow tree, or from planks as in cask-making. Earthenware ones are worst, both freezing in winter and boiling in summer. §7 Within the enclosure which I recommended, bases should be fashioned three feet high, covered with crushed-tile plaster and smoothed with stucco-work to prevent the damage caused by lizards and other creatures that have the habit of creeping in. On these bases the hives should be placed in such a way that they cannot be penetrated by rain, and separated from each other by little gaps. Only a narrow entrance should admit the swarms, because of the harmful effects of cold and heat. Naturally, colder winds should be blocked by a high wall, which can reflect the sun's warmth onto the place while protecting the hives. §8 All the

---

62. The smoke from burning crabs was thought beneficial in the vineyards against blight (Pliny 18.293, *Geoponika* 5.33.2).

entrances should face the winter sun. There should be two or three of them in one cork hive, of a size not to exceed the dimensions of a bee. For in this way vermin will be thwarted by the narrow entrance, or, if they want to ambush the bees as they fly out, they can make use of another exit, since it will be available.

38. *Purchasing bees*

§1 If bees need to be purchased, we must make sure that full hives are obtained. This is ascertained either by inspection or by the loudness of their buzzing or the numbers of the swarm as it comes and goes. Best from a neighbouring region rather than a distant one, in case they are afflicted by a strange climate. But if they have to be brought from further afield, they should be carried on people's necks at night;[63] and we should not put the hives in place or open them until evening approaches. §2 Then we should keep watch for three days, in case the whole swarm leaves the entrance; for this is a sign that they are planning to make off. We shall discuss steps against this, and other topics, each in the relevant month.[64] It is thought, however, that they will not make off if we smear the mouths of their cells with the dung of a first-born calf.

39. *The bath-house*

§1 It is not inappropriate, if the water supply permits, for the head of household to consider building a bath, something that contributes greatly to pleasure and to health. We shall establish the bath, then, in an area which will be warm, on a spot raised above moisture, so there is no wet ground near the furnace to cool the place down. We shall give it lights on the south side and facing west in winter, so it is cheered and brightened all day by the sunlight. §2 As for the raised floors of the rooms, you will make them as follows. First you pave the area with two-foot bricks; the pavement should slope towards the furnace so that if you roll a ball onto it, it cannot stop there but runs back to the furnace. The result will be that the flames, seeking

---

63. They are carried by people rather than in carts to minimize jolting, and at night because they are quiescent.
64. Preventing departure: 7.7. Other topics: 4.15, 5.7, 6.10, 7.7, 9.7, 11.13, 12.8.

the higher area, make the rooms warmer. On this pavement piers should be built, of bricks with kneaded clay and hair; they should be a foot and a half from each other, and two and a half feet high. Resting on these piers, two-foot bricks should be set in a double layer; over these a floor of crushed-tile mortar should be poured, and then, if you have a supply, marble should be laid.

§3 We shall set a water-heater of lead, with a shallow copper pan beneath it, outside, between the locations of the bath-tubs,[65] with the furnace below it. A cold-water pipe should lead to this water-heater, and a pipe of similar size should run from it to the bath-tub; this pipe will draw an amount of hot water into the bath equal to the cold water brought to the heater by the other pipe. The rooms should be arranged so that they are not square, but for example if they are 15 feet long, they should be 10 feet wide; for the heat will roll around more vigorously in a narrow space. The shape of the bath-tubs can be according to individual preference.

§4 Rooms with swimming-pools should be lit from the north in summer bath-houses, and from the south in winter baths. If possible, the baths should be placed so that all the grey water from them runs through the gardens.

In bath-houses ceilings of waterproof mortar are stronger. Those made of planks are supported by horizontal iron rods and iron arches.[66] §5 But if you do not want to use planks, on the arches and rods you will place two-foot bricks, fixed with iron clamps and mortared together with kneaded clay and hair, and then coat the undersurface with crushed-tile plaster; then you will decorate it with bright stucco-work. If we are keen on economy, we can also place winter living-quarters on top of the bath-house; in this way we provide warmth from below for the living area, and make additional use of the foundations.

65. i.e. outside the bath-house wall, and not directly opposite any of the tubs (which are of course inside) but between them. See Figure 5 for an illustration. The purpose of the copper pan is to protect the water heater from direct heat, since lead melts at a relatively low temperature.
66. The planks, like the bricks given as an alternative, would be coated with layers of plaster.

40. *Caulks for hot and cold water*

§1 Since we are talking about baths, it is convenient to know about the caulks for hot and cold water, so that if a crack ever occurs in the material of the bathtubs, it can speedily be mended. The composition of those for hot water is as follows. Take hard pitch and white wax in equal amounts by weight; tow; liquid pitch weighing half of the whole mixture; pulverized earthenware and flower of lime.[67] Mix and combine these all together, and insert carefully into the joints. Alternatively, pound ammoniac gum[68] that has been melted, figs, tow and liquid pitch in a mortar and smear it on the joints. §2 Alternatively, melt both ammoniac and sulphur, and smear this on the joints or pour it into them. Again, melt hard pitch and white wax plus ammoniac and sulphur together, smear on the joints, and run over them all with a cauterizing iron. Again, smear flower of lime mixed with oil into the joints, and take care not to have water introduced soon. §3 Alternatively, mix powdered lime into bull's blood and oil, and use this to cover the cracks in the joints. Again, you can pound together figs, hard pitch, and dry oyster shells, and carefully smear the joints with all these materials.

Again, cold-water caulks. Take ox blood, flower of lime, iron slag; pound everything together with a pestle until it is like wax ointment, and carefully smear it on. Again, melted tallow mixed with sieved ashes will stop water escaping through the cracks, if smeared onto them.

41. *The mill*

If there is a good supply of water, the run-off from the bath should be guided to the mill as well.[69] So, by setting up water-driven millstones, the grain can be broken without labour by beasts or men.

---

67. Powdered lime.
68. So called because produced near (H)ammonium in North Africa.
69. As well as to the gardens (1.39.4).

42. *Tools for countryfolk*

§1 We should also prepare the tools that are necessary for the country:

- ∞ ploughs, either simple or (if level terrain permits) eared,[70] so that a raised ridge can lift crops above standing water in winter
- ∞ two-pronged hoes
- ∞ picks
- ∞ pruning hooks for use on tree or vine
- ∞ §2 reaping and hay-cutting hooks
- ∞ mattocks
- ∞ 'wolves', i.e. small saws with handles, longer or shorter (up to a cubit in length), which can easily be inserted (as an ordinary saw cannot) in trimming the trunk of a tree or vine
- ∞ 'needles'[71] for planting cuttings in dug-over ground
- ∞ hooks that are sharpened on the back and crescent-shaped
- ∞ short curved knives, which make it easier to cut away dry or protruding side-shoots on young trees
- ∞ §3 very small tubular[72] hooks, which we use to cut bracken
- ∞ smaller handsaws
- ∞ spades
- ∞ grubbing-tools, with which we attack brambles
- ∞ axes, plain or with picks[73]
- ∞ hoes, either plain or two-pronged or with mattock blades on the back
- ∞ drag-hoes
- ∞ cauterizing irons, castration and shearing tools and those that pertain to animal medicine
- ∞ §4 also leather tunics with hoods, and leggings and sleeves made of leather, which can serve a double purpose, for farmwork and hunting, either in the woods or amongst brambles.

70. The 'ears' pushed up the soil on each side of the ploughshare, thus creating ridges.
71. i.e. dibbles or dibbers.
72. Meaning uncertain: perhaps 'with tubular sockets' for fastening to a handle (Hamblenne), or 'with tubular handles' for lightness (Schmitt).
73. Those with picks would have a vertical blade on one side and a horizontal pickaxe blade on the other.

Book 2: January

Now that I have covered the category of general advice, I shall describe the tasks that belong to each individual month. Let us make a start, then, with the month of January.

Field Work

1. *Ablaqueating vines*

In temperate regions January is the time to ablaqueate vines (a process Italians call *excodicare*), i.e. thoroughly open up the earth with a pick around the stem (*codex*) of the vine, clean everything up and create a kind of pool,[1] so it can be stimulated by the sun's warmth and the showers.

2. *Meadows*

In places that are sunny and thin-soiled or arid, now is the time to clean up the meadows and keep livestock off them.[2]

3. *Ploughing*

§1 Rich dry fields can be first-ploughed now and made ready. It is better to yoke the oxen by the neck rather than the head.[3] When

---

1. i.e. a hollow where rainwater and surface water can pool and be taken up by the vine's roots.
2. To promote a good hay crop.
3. Pliny 8.179 notes that small-bodied Alpine cattle were yoked not by the neck but the head (presumably by attaching the yoke to the horns).

they come to the turn, the ploughman should hold them back and push the yoke forward, to cool their necks. The furrow in ploughing should be no longer than 120 feet. We must ensure that no earth is left undisturbed between the furrows. All clods should be broken up with picks. §2 You will know the earth has been uniformly disturbed, if you drive a long pole crosswise through the furrows. Doing this quite often will deter the ploughboys from negligence in this regard.

We must ensure that a muddy field is not ploughed, nor one wetted on the surface with a light rain shower after a long drought, as often happens. §3 For land that is worked at the outset when muddy cannot then be worked, it is said, for a whole year; while land that is lightly moistened above but dry beneath, if ploughed in that condition, is alleged to become sterile for three years. And so it is a moderately dampened field, neither muddy nor dry, that should be first-ploughed.

If there is a hill, it should be furrowed horizontally around its flanks. This pattern should be preserved when the land is seeded.[4]

Sowing of crops

4. If the winter is mild, we should sow *Galatian barley,*[5] which is heavy and white, around 13th January in temperate places. A juger will be covered by eight *modii*.

5. *Small chickpea*[6] is sown this month in a fertile location and in damp weather. Three *modii* cover a juger. But this crop rarely does well, since it is spoilt if caught in flower by a south wind or a drought, conditions that almost inevitably occur at that time.

6. At the last of this month *vetch* is sown – for the purpose of collecting seed, not for cutting fodder.[7] Six *modii* fill a juger. It should be sown in land that has been first-ploughed, after the second or third hour, once there is no dew, which it cannot stand. But it must be

---

4. A disagreement with Columella, who says subsequent ploughings on hills should be at an oblique angle to the first, to break up the ground more thoroughly (2.4.10).

5. This was a superior variety of barley, used mixed with wheat for human consumption according to Columella 2.9.16.

6. Identified by some authorities as *Lathyrus sativus*, chickling vetch.

7. Vetch grown for fodder is sown in September (10.8).

covered up straightway before nightfall; for if it remains uncovered, it is spoilt by the dampness of night. Take care not to sow it before the twenty-fifth day of the moon, since slugs attack it if it is sown so.

7. In Italy we shall sow *fenugreek*, for the purpose of collecting seed, at the end of January and the beginning of February. Six *modii* will suffice for a juger. The ploughing should be close but not deep, for if the seed is buried more than four fingers deep, it has difficulty in growing. For that reason some people first-plough the ground with the smallest size of plough before seeding, and cover the seed straightway with hoes.[8]

8. *Bitter vetch* can be sown this month too,[9] at the end of the month, in a dry place that has lean soil. Five *modii* are sown to a juger.

9. *Hoeing the crops*

§1 On clear dry days this month, in the absence of frost, the grain crops should be hoed. Most people say this task should not be done, on the grounds that the roots are uncovered or cut, and killed by subsequent cold weather. My opinion is that it should be done only in places that are full of weeds. Wheat and emmer are hoed when four-leaved, barley when five-leaved, beans and legumes when they are four fingers above ground. §2 But lupine, which has only one root, is killed if hoed – and it does not require hoeing, since it overpowers the weeds without the grower's help. Beans, on the other hand, will thrive if hoed twice; they will return a large yield in quantity and size, so that in filling a measuring-vessel they do almost as well when shelled as when whole. If you hoe the crops when dry, you have given them some help against the rust as well. Barley in particular should be hoed when dry.

---

8. Again there is another sowing, for fodder, in September (10.8). The ploughing is done before seeding, to prepare the ground.

9. It can also be sown in February (3.7) and in autumn (11.1).

Planting Vines

10. *Ground preparation*

§1 Now is the time for ground preparation. This is done in one of three ways: digging the ground over completely, or digging trenches, or digging holes. Where a field has not been cleaned, the ground should be completely dug over, in order to clear the area of tree trunks and roots of bracken or noxious weeds. But where there are clean new fields, we can prepare the ground by digging holes or trenches. Trenches are better, because then the moisture is spread virtually throughout the prepared ground.

§2 Trenches, then, are made of the length you have decided on for the planting bed, and two and a half or three feet in width, so that two diggers working as a pair can tackle with their hoes a strip marked out with a line, and three or two and a half feet in depth.[10] Next, if the vineyard is to be cultivated by hand, we leave the same width of undug soil, and then a second trench is sunk. But if the vineyards are to be ploughed,[11] we shall leave a space of five or six feet which is not to be dug between the trenches.

§3 If holes are the choice, we shall make them three feet deep, two and a half feet wide, and three feet long. According to whether the vineyards are to be cultivated by diggers or by oxen, we should keep the same distances as were prescribed between trenches. Holes should not be dug deeper than three feet, to prevent the cuttings that we plant from being troubled by cold. The sides of the holes should be cut evenly, to prevent a vine set at an angle from being wounded if tools penetrate deeply when the digger bears down on them.

§4 In prepared ground that is to be entirely turned over, all the earth needs to be dug to a depth of two and a half or three feet. Careful attention will be paid to ensure that the digger does not

10. Two men working as a team with hoes are shown on the mosaic illustrated at White *Agricultural Implements,* plate 3 (= *Roman Farming,* plate 26). Hoes seem inappropriate for digging to the depth envisaged here; possibly they were used to break up the soil, which was then dug out with spades. (For this system see Pliny 17.159 and White, *Roman Farming,* plate 28.)
11. i.e. the strips of land between the rows of vines, whether ploughed for inter-cultivation or just to control weeds and open the soil.

deceitfully cover over undug soil. The overseer should repeatedly check this throughout the areas being dug, using a rod on which the level of the aforesaid depth is marked. He must ensure that all roots and refuse, especially from brambles and bracken, are thrown up to the surface. This care needs to be exercised in all regions and however the land is situated.

Measurement of area

11. We shall make the planting beds in accordance with the owner's inclination or the requirements of the place, covering a whole juger or half or a quarter bed, which consists of a fourth of a juger in square footage.

12. In a square planting bed covering one juger, the measurement of the prepared ground is 180 feet on each straight side; when multiplied this will yield 324 10-foot square units across the whole area. Using this figure, you will divide up all the ground you want to prepare. For 18 10-foot lengths multiplied 18 times will yield 324. This example will show you how to measure a larger or smaller field.[12]

13. *Soil, climate, location and ground preparation*

§1 The soil for planting a vineyard should be neither too dense nor loose, but closer to loose; neither thin nor very rich, but closest to rich; neither flat nor steep, but rather on raised level ground; neither dry nor boggy, but moderately moist; neither salty nor bitter, since that fault spoils the wine's flavour and makes it sharp. §2 The climate should be of middling character, but warm rather than cold, and dry rather than too rainy; above all else, however, the vine fears storms and winds.

---

12. In vine-planting careful measurement of the area was important to calculate the number of man-days required for hand-digging (a major expense) and the number of vines needed for planting, which takes place next month (3.9).

The basic unit of Roman land measurement was the juger (*iugerum*), roughly two-thirds of an acre. It was traditionally defined as a rectangle measuring 240 x 120 Roman feet, i.e. 28,800 square feet. But when defined as a square it was said to have 180' on a side (180 being the mid-point in length between 240 and 120), which yields 32,400 sq. ft. as Palladius says. Palladius uses the square juger again in calculating the number of vines needed (3.9.9–10).

In preparing the ground we should choose unbroken land, especially if wooded. The least desirable option is a place where there were once old vineyards. If necessity drives us to this, it must first be worked over with the plough many times: in this way, when the roots of the former vineyard have been destroyed and all its mouldy decayed matter ejected, new vines can be brought in more safely. §3 Tufa and other harder matter, when softened by frost and sun, produce very fine vineyards since the roots are cooled in summer and moisture is retained; so do free gravel and pebbly ground and loose stones, as long as all these are combined with rich clods of soil; and broken rock with earth above, since it is cold and moisture-retaining, prevents the roots from going thirsty in summer. Also places onto which earth washes down from mountain heights, or valleys enriched by the accumulation from rivers – but only where there is no possibility of the area being harmful through frost or mists. §4 Ground containing argil is favourable, but pure argil is seriously antipathetic, and the other types of ground that I mentioned in the general section (1.5). For a place that produces miserable vegetation is thereby proved to be boggy or salty or bitter or thirsty and dry. §5 Black or red grit is usable, as long as it has robust soil mixed in with it. *Carbunculus*[13] makes for thin vines, unless it is manured. In red clay they have difficulty in taking root, though they are nourished by it later. But this kind of ground is antipathetic to being worked, since just a little moisture or sun makes it too wet or too hard. At any rate, the most usable soil is one that will keep a state of balance between all extremes, and which is much closer to being loose-knit than hard-packed.

§6 The orientation of the vineyard should be southerly in cold places, northerly in hot ones, easterly in warm ones – but only if the region does not suffer from destructive south or east winds. If it has this defect, we shall do better to orient the vineyard to face the north or west wind.

§7 The place that is to be prepared should first be cleared of obstacles and of all trees, so that the earth once dug over will not be packed hard by constant trampling later.[14] If it is level ground, it

13. Meaning uncertain: apparently a form of sandstone.
14. i.e. if workers come in to remove trees, etc. after digging-over has started.

should be dug to two and a half feet; if a slope, to three; if a steep hillside, to four, to prevent the earth slipping down too quickly; if valley land, just two feet. But boggy land that disgorges moisture if dug deeply, as in the territory of Ravenna, is dug no further than a foot and a half.[15] §8 I have learnt by repeated tests that vines grow better if they are planted in ground that has just been dug, or not long before, while the aerated ground has not yet subsided or regained its solidity. I have also tested this and found it to be the case in making trenches and holes, especially where the earth is ordinary.

14. Gardens

Lettuce

§1 Lettuce should be sown in January or December, so the seedlings can be planted out in February. Likewise it is sown in February, for planting out in April. But it can certainly be sown successfully throughout the year, if the place is fertile, manured and irrigated. Before it is planted we should cut back its roots to an even length and wipe them with liquid dung; alternatively, if they have already been planted, you can uncover their roots and manure them. §2 They love a soil that is worked over, rich, moist and manured. Weeds between them should be pulled by hand, not with a hoe. They become broader if planted well-spaced, or if, when they begin to produce a stalk, it is lightly cut and weight is applied with a clod of soil or a tile. §3 They are thought to grow blanched if you frequently scatter sand from river or shore on the hearts, and tie them with their leaves bunched together. If lettuce quickly becomes tough through some fault of the place or season or seed, the plant will gain in tenderness if pulled out and planted anew.

Moreover it will grow flavoured by several other seedlings, if you delicately hollow out a goat dropping with an awl, and place in it one

---

15. §7 is an odd reversion to the topic of digging-over (already covered in chapter 10 above), possibly caused by plural sources. Here Palladius follows precisely the recommendations of Columella 3.13.6–8; in chapter 10, by contrast, his recommendations often differ from those of Columella 3.13.2–5, and he may have been using a supplementary source.

seed each of lettuce, cress, basil, rocket and radish, then cover the pellet in dung and drop it in a shallow hole in very well-cultivated earth. The radish pushes into the root; the other seeds break the surface as the lettuce simultaneously absorbs them, preserving the flavour of each. §4 Others achieve the same thing this way; they pull up a lettuce, pluck the leaves that are close to the roots, make holes in these tiers with a twig, drop the above-mentioned seeds in them (except the radish), and smear with dung. Planted again like this, the lettuce will be surrounded by the stems growing from the aforementioned seeds.

Lettuce (*lactuca*) is so named because it has copious quantities of milk (*lactis*).

Cress
§5 People agree that cress may be planted this month and indeed at any time, in whatever location and climate you please. It does not want manure; although it loves moisture, it is not troubled by lack of it. If sown with lettuce, it is said to germinate exceptionally well.

Other sowings
Do not hesitate to sow *rocket* both now and in whatever months and locations you wish. Also this month, and throughout the year, *cabbage* can be sown, but it is better done in other months as prescribed.[16] This will also a good month for sowing *garlic* and *ulpicum*;[17] for garlic, light-coloured soil will be beneficial.

15. Fruit Trees

Service trees
§1 Service trees are sown with excellent results in January, February and March in cold places, but in hot places in October and November, by planting the actual ripe fruits in a nursery bed. I have seen many trees, productive both in growth and in bearing, that started spontaneously from the fruits. §2 If anyone wishes to plant the

16. February and March (3.24.5, 4.9.5).
17. A large-headed variety of garlic.

seedlings, he also has that option, so long as he plants them out in November in hot places, in January or February in temperate places, and as March declines in cold places. The tree loves places that are moist, montane, and verging on cold, and soil that is very rich; this last is reliably indicated if the tree grows plentifully throughout an area. The seedling should be transplanted when quite robust; it wants a fairly deep hole and generous space, so it can be buffeted by frequent winds (something which is particularly beneficial for it), and hence reach a good size.

§3 If they suffer from damaging worms, which are red and hairy on this tree and regularly go after the inner pith, we should remove some of these without damaging the tree, and burn them on a fire nearby. By all accounts they are either driven off or killed in this way. If it begins to bear less, a pine wedge should be inserted in its roots, or else a channel should be dug around the foot of the tree and levelled off by piling ashes in it.

April is the month for grafting service trees, onto themselves or quince or whitethorn, either in the trunk or in the bark.[18]

§4 Service-berries are preserved as follows. They are picked when quite hard and set aside; then, when they begin to soften, they are packed in little earthenware jars right to the brim. These are sealed on top with gypsum and buried upside-down in a two-foot hole in a dry open-air location, with the earth above trodden down quite firmly. Alternatively they are cut in pieces and dried in the sun, and kept in containers till winter. §5 When we want to use them, they freshen up with an agreeable flavour if steeped in boiling water. Some folk pick them green with the stalks and hang them up in shady dry places. People also say that wine and vinegar can be made from ripe service-berries in the same way as from pears.[19] Others state that service-berries can be preserved for a long time in *sapa*.

Almonds

§6 The almond is started in January or February, and in hot places in October or November, using either seeds or slips taken from the

---

18. For these grafting methods see 3.17. Whitethorn (*spina alba*) is probably a species of hawthorn.

19. Recipes for pear wine and vinegar are given at 3.25.11.

root of a larger tree. For this species of tree, nothing is more effective than making a nursery bed. Accordingly we shall dig over a plot to a depth of one and a half feet, and plant the almonds there no more than four fingers deep,[20] embedding their points in the ground at a distance of two feet from each other. §7 They love ground that is hard, dry and pebbly, and a very hot climate, since they flower early. The trees should be positioned so as to face south. Once they have grown up in the nursery bed, we shall leave sufficient plants there for the space, and transplant the others in the month of February. The almonds we collect for planting should be fresh and large; the day before planting we should soak them in honeywater – with a high proportion of water, to prevent the germ being killed by the stinging effect of too much honey. §8 Others soak these nuts first for three days in liquid dung, then leave them for a day and a night in honeywater, which can have only a suspicion of sweetness. When we plant out the almonds in the nursery bed, we should water them thrice a month if dry weather supervenes, and dig round them frequently to clear them of weeds that spring up. The soil of the nursery bed should have manure mixed into it. Fifteen or twenty feet should be allowed as a sufficient distance between trees.

§9 Pruning should take place in November, with the purpose of removing superfluous, dry and crowded branches. They should be protected from livestock, because they become bitter if browsed. They should not be dug around when flowering, since that results in their flowers being dropped. They bear more as they age. If a tree is infertile, we should bore a hole in the root and insert a wedge of pinewood, or a stone, in such a way that it will be enclosed as the bark covers it. §10 In cold places, where there is fear of frost, Martialis says the following remedy is applied:[21] before flowering the roots are laid bare and tiny white stones mixed with sand are piled on them; then, when it seems safe and timely for them to blossom, the stones are dug out again and removed. §11 The almond tree will form soft

20. As in measuring whisky today, 'fingers' refers to their width not length. A finger is conventionally $\frac{1}{16}$ of a Roman foot. See 'Measures' at the end of the Introduction.

21. Though most of Gargilius Martialis' work on fruit trees is lost, we have a surviving fragment that discusses quince, peach, almond and chestnut: Martialis does indeed say just what Palladius reports here.

nuts, he says, if the roots are ablaqueated before flowering and hot water is poured on them for several days. Bitter trees become sweet if you dig around the trunk and make a cavity in it three fingers from the root, through which the tree can sweat out the noxious moisture; or if a hole is bored through the middle of the trunk and a wooden wedge smeared with honey is forced into it; or if you pour pig manure around the roots.

§12 Almonds declare their ripeness for picking when they divest themselves of their outer coating. They keep for a long time without human care. If the shell resists being removed, it will loosen immediately if the nuts are covered with chaff. Furthermore, if we wash them in seawater or salt water once shelled, they not only become bright but also last very well.[22]

The almond is grafted in December or mid-January, but in cold places in February too, as long as you store the scions before they blossom.[23] Those taken from the top of the tree are effective. It is grafted both under the bark and in the trunk. It is grafted onto itself or onto peach.

§13 The Greeks state that almonds will grow with writing on them if you open a shell, remove a healthy nut, write something on it, close it up again, cover it in mud and pig manure, and plant it.[24]

Walnuts

§14 We shall sow the walnut at the end of January or in February. It loves places that are montane, moist, cold, and mostly stony. However, it can be cultivated in temperate places too with the aid of moisture. It should be sown by means of the nuts, in the same way that almonds are sown and in the same months. If you set any out in November, you will dry them somewhat in the sun, so the harmful secretion of moisture can be drawn out. §15 But those you intend to plant in January or February you will steep the previous

22. Presumably the salt water blanches the almonds (a process done nowadays with boiling water). In Palladius' source Gargilius the chaff is used to loosen the hulls, not the shells, but the sentence does invite misunderstanding.
23. If scions are cut in winter and stored in a cool damp place, they remain dormant but fresh until needed.
24. A similar trick supposedly produces peaches with writing on them (12.7.3). Democritus is credited in the *Geoponika* as the source for both tricks.

day in plain water. We shall plant them horizontally, so that the flank, i.e. the half-shell, is embedded in the ground. As for the point, we shall orient it to the north when planting the nut. A stone or crock should be placed beneath, so it will not just make one root but form a root-cluster when it hits this obstacle.

It will be more prolific if transplanted several times. In cold places it should be transplanted at two years, in hot places at three. §16 In this species you should not cut back the transplant's roots, as is our practice with other trees. The bottom of the plant should be dipped in ox dung. But it is better to sprinkle ash in the holes, to prevent burning by the heat of the manure, and ash is believed to induce either softness in the shell or abundance in the fruit. §17 It appreciates deep holes, in keeping with the tree's large size, and it requires wider spacing, since it will harm nearby trees, even of its own species, by the droplets falling from its leaves.

It should occasionally be dug around, so as not to become hollow through the weakness of age. A long groove should be gouged out from the top of the trunk to the bottom: in this way the beneficial effects of sun and wind can harden those areas that were starting to rot. §18 If a walnut is hard or gnarled, the bark needs to be cut around, to draw off the harmful moisture. Some people cut back the ends of the roots, others bore a hole in the root and push into it a peg of boxwood or a nail of copper or iron. If you want to make it Tarentine, you will have to take just the flesh of the nut, wrap it in wool because of ants, and bury it in the nursery bed.[25] If you want to change a bearing tree into this type, you will water it with lye thrice a month for a whole year.[26] The shedding of the nut's outer covering is a sign of ripeness, and that is the condition in which it should be planted.

§19 Walnuts are preserved by being buried in chaff or sand or their own dried leaves, or enclosed in a box made of their own wood, or mixed with onions, in which case they return the favour by

25. Columella says that a Tarentine version (which he does not explain) of an almond or hazel tree can be produced by planting the nut without its shell (5.10.14). According to *Geoponika* 10.66, if you plant a walnut or almond without its shell, the resulting tree will bear shell-less nuts.
26. Since lye was made from wood-ash, it was naturally thought to have the same softening effect as ash placed in the planting-hole.

removing the onions' sharpness. Martialis says[27] that to his personal knowledge green nuts, simply freed of their shells, are plunged in honey and after a year are still green, while the honey itself becomes so curative that a draught made from it cures the breathing passages and the throat.

It is grafted, according to most authorities, in February, onto arbutus (preferably in the trunk); also, according to some, onto plum or onto itself.

Other fruits, mostly planted in other months
§20 This month the *azarole* is grafted onto quince. In temperate places *peach* stones are planted now, and peach is grafted onto itself or almond or plum – but we shall graft the Armenian and early-ripening varieties only onto plum. Now too *plum* should be grafted, before it starts exuding gum, onto itself or peach. This will also be a good time to graft *wild cherry*.

Miscellaneous

16. This month, as Columella says, early lambs and all animals, larger and smaller, should be marked with an emblem.[28] This is the appropriate time for the process of preserving bacon, salted sea-urchins, turnips and hams.

Making oil and wine from shrubs

17. Myrtle oil
This month you will produce oil with myrtle berries, as follows. [*Material lost*][29] You add an ounce of leaves per *libra* of oil, and a

---

27. Not in his surviving work on medicinal uses of fruits and vegetables: probably, therefore, in his largely lost work on fruit trees.
28. Columella 11.2.14. Early lambs are those born in autumn, when most lambs were born in antiquity. Late-born lambs are to be marked in April (Palladius 5.6).
29. Two different procedures for making myrtle oil have become conflated here, probably through omission of material by a copyist. The first method is making oil from myrtle berries; the second is flavouring olive oil with myrtle leaves. Both procedures are described by Pliny (15.27), but Palladius provides more detail than Pliny on the second.

hemina of old astringent wine per 10 ounces of leaves, and boil it with the oil. The reason the leaves are sprinkled with wine is to prevent their drying out before they are fully cooked.

18. *Myrtle wine*
Again, from the same plant's berries you can produce myrtle wine, as follows. In 10 urban *sextarii* of old wine you put crushed myrtle berries, 3 urban *sextarii*,[30] which should be infused for 19 days. Then after squeezing the myrtle berries you strain it; in the wine you place half a *scripulus* of saffron and one *scripulus* of *folium*, and blend all this with 10 *librae* of the best honey.[31]

19. *Laurel oil*
Again, oil from laurel berries will be produced in the following way. You will take laurel berries, as many as possible, plump with ripeness, and seethe them in hot water. When they have boiled for a long time, the oil they release will float in a layer on top; you will gather this gently with feathers and transfer it to vessels.

20. *Mastic oil*
The time is also ripe for production of mastic oil, which is made as follows. You collect ripe mastic berries, as many as possible, and let them stand in a pile for a day and a night. Next you place a basket full of the berries on top of any kind of container, add hot water, tread and press them. Then, from the liquid that flows out, the mastic oil floating on top will be collected just like laurel oil. But to prevent it from cooling and congealing, remember to pour on more hot water frequently.

30. 'Urban' here = 'Roman'. In Greek writers, 'Roman' or 'Italian' denotes that the *sextarius* is a non-Greek measure. Consequently the present phrase, and 'Italian *sextarii*' at 11.14.14, probably originated from a Greek source.
31. Myrtle wine was used for various medicinal purposes. The berries were readily available, and several recipes survive from antiquity; none of them, however, includes saffron or *folium*, so Palladius is apparently following a different source. *Folium* refers to a highly perfumed leaf, probably patchouli or nard. A *scripulus* is $\frac{1}{24}$ of an ounce, i.e. 1.14 g.

Et cetera

21. Laying hens become productive again this month after their winter respite, and one starts putting eggs under them for rearing of chicks.

22. This month too wood should be cut for building, on a waning moon, and vine-stakes or posts should be fashioned.

23. *Hours*[32]

This month tallies with December in the length of the hours, whose measurement is calculated thus:

Hour:	1	2	3	4	5	6	7	8	9	10	11
Feet:	29	19	15	12	10	8	10	12	15	19	29

32. Palladius provides a corresponding section on time-reckoning at the end of each month. The measurement in feet is the length of a person's shadow, enabling him to act as a human sundial and estimate the time wherever he might be on the farm. The Roman day from sunrise to sunset was divided into 12 equal hours (with the result that the length of the hours was relative to the time of year).

Book 3: February

Field Work

1. *Meadows*

In temperate places one starts to protect pastures this month. First, if they are thin, one should drench them by scattering manure, which should be broadcast on a waxing moon. The fresher it is, the more effective it will be in nourishing the grass. It should be spread on higher ground, so the liquid from it can flow through the whole area.

2. *Ploughing*

As early as this month, in warm areas or if the weather is mild and dry, you can first-plough hills that have rich soil.

Sowing crops

3. This month the whole range of *three-month crops* should be sown.[1]

4. This month too you will sow *lentil*, in soil that is thin and loose or even rich, but above all dry, because it is spoilt by lushness and moisture. The proper time for sowing is up to the twelfth day of the moon. So that it will come up quickly and grow well, it should first be mixed with dry dung; after sitting in this mixture for four or five days, it is then sown. The seed of one *modius* will fill a juger.

---

1. These are cereal crops adapted to ripen in three months from sowing; see 1.6.16.

This month too *small chickpea* is sown, in the location and manner I described earlier (2.5).

5. At the end of this month you sow *hemp*, in land that is rich, manured, watered or level and moist, and worked to a good depth. Six seeds of this crop are planted per square foot.

6. *Preparing the fields for alfalfa*

Now the land that is going to receive alfalfa (a crop whose nature we shall discuss at sowing time [5.1]) should be second-ploughed, cleared of stones and harrowed thoroughly. And around the end of the month, once the soil has been worked as in gardens, seedbeds should be formed 10 feet wide and 50 feet long, such that water can be supplied to them and they can easily be weeded from each side. Then, after a topping of aged manure, they are reserved in readiness for April.

7. *Bitter vetch*

It is still possible to sow bitter vetch throughout this month; but it should not be sown in March, lest it harm livestock that feed on it, and make oxen go mad.[2]

8. *Urine as fertilizer*

At this period, if aged urine is poured on fruit trees and vines, it improves the fruit both in abundance and in shape.[3] It will be helpful to mix in unsalted *amurca*, especially for olives – but this on colder days, before the hot weather begins.

Galatian barley

Now too Galatian barley, which is heavy and white, will be sown in cold places at the end of the month.

---

2. It was believed to be noxious if sown in March, but not if sown earlier (Pliny 18.139).

3. Columella recommends ageing the urine for six months (2.14.2).

9. Vines

Choice of varieties

§1 This month dug-over ground of all kinds, or trenches or holes,[4] should be filled with vines. The nature of the vine tolerates any climate and soil, if its varieties are appropriately employed. §2 So in flat ground you will establish a variety of vine that tolerates mists and frosts; in hills, one that endures drought and winds; in rich land, slender varieties that are less prolific; in lean land, productive substantial ones; in heavy land, strong leafy ones; in cold misty land, varieties that forestall winter by maturing early, or that have hard grapes and can flower with safety in gloomy weather; in a windy situation, firm-rooted types; in a hot situation, those with softer moist grapes; in a dry situation, those that cannot endure rains. In short, we should choose varieties which, as their weaknesses tell us, desire locations that are the very opposite of those in which they were not able to survive. Of course a calm region with fine weather will safely support all varieties.

§3 It is not pertinent to list the varieties of vines. But it is well known that larger grapes of fine appearance, hard-skinned and fairly dry, are best kept for the table, while highly productive ones, more tender-skinned and fine-flavoured, and especially those that shed their flowers early, should be kept for wine-making. The location changes the nature of most vines. §4 Only Aminean vines produce the finest wines wherever they are. They will tolerate a hot situation better than a cold one. They cannot be moved from rich to lean soil, except with the help of manure. There are two kinds, the larger and the smaller; the smaller sheds its flowers better and earlier, with shorter internodes and smaller grapes. It needs rich soil if attached to a tree,[5] but middling soil if cultivated in rows. It scorns rain and winds, whereas the greater is often spoilt in flower. Apian vines are also outstanding. It is sufficient to have mentioned these varieties.

§5 A painstaking man should choose varieties that have been tested, and commit them to the kind of ground that resembles that

---

4. These are the three planting situations described in 2.10.
5. i.e. in a vineyard where the vines are supported by trees, as described in chapter 10 below.

from which they are taken; in this way each will maintain its own virtues. But it will be better to shift a vine or tree from thin ground to rich. For if they are moved from rich ground to thin soil, they cannot be productive.

Selection and handling of cuttings

The cuttings that we plant should be chosen from the middle of the vine, not from the top or bottom; they should project a distance of five or six buds from the old wood, since cuttings transferred from such places do not easily degenerate. §6 Mind, they should be taken from a productive vine. And we should not consider as fruitful those branches that yield one or two clusters of grapes, but those that are weighed down with a great abundance. For a fertile vine can have within it even more fertile growth. Another sign of fertility will be if it puts out fruit from somewhere on its hard wood, or if the little branches rising from the base are filled with produce. §7 This should be noted by putting markers in place during the vintage. A new vine-shoot chosen for planting should not have any of the hard old branch-wood, since a shoot is often spoilt by the decay of such material. We should reject whips and shoots from high up: though they grew in a favourable place, they have been found to lack fruitfulness. §8 A stock-shoot that grows from hard wood, even if it has produced, must not be planted as if it were fruitful: for in its original position it is made fertile by the mother-vine, but when transplanted it retains the failing of sterility which it acquired through the circumstances of its growth.

When the cutting is planted, its top should not be twisted or treated roughly in any way: otherwise the more fertile part is buried while the virtually sterile part is left above ground. In addition, twisting is itself rough handling, and the part from which the root is expected to grow must not be subjected to any injury with which it would be forced to struggle before it can take hold.[6] Vines should be planted on calm warm days, and care should be taken that the cuttings are not parched by sun or wind, but are either planted straightway or kept covered.

6. An earlier authority, Julius Atticus, had recommended bending the cuttings so that the part below ground would be horizontal and the part above ground vertical (Columella 3.18.2–6); Palladius rejects this advice.

Spacing of cuttings

§9 Planting of vines should take place this month and throughout the remainder of the spring in cold rainy areas, on rich flat ground and in moist regions. The length of a cutting should be one cubit. Where the nature of the ground is rich, we shall leave larger spaces between vines; where it is thin, tight spaces. So some people, when planting vines in soil that has been completely dug over, allow three feet in each direction between the vines. But by this spacing method 3,600 cuttings will be planted in a plot measuring one juger. §10 If, however, you decide to leave two and a half feet between vines, 5,476 vines will be planted in a plot of the same size.[7] In planting we shall observe the following sequence. We shall flag a string with white markers (or any kind), keeping the distances we have decided to maintain; then we shall stretch the string across the bed, and push twigs or canes into the spots where each vine is to grow. In this way the whole area of the bed will be filled with twigs corresponding to the number of future vines, and the man who is going to do the planting will set in the cuttings, which have been laid beside the twigs, without any error.

Diversifying plantings

§11 Furthermore, one should avoid planting the whole of the dug-over area with a single variety of vines, for fear that a year that is unfavourable to that variety should wipe out the hope of the whole vintage. For that reason we shall plant cuttings of four or five outstanding varieties. It will be extremely useful to separate the varieties in different beds, and divide them with parallel walkways, unless the difficulty of that task is a deterrent. But if we can graft old vineyards with scions of particular varieties, each in a separate bed, we shall actually find it easy to achieve this method of cultivation, which is attractive-looking and functional. By this method we shall be able to have flowering and ripening times, which differ between varieties of vines, occurring seasonally for each variety. §12 (No small loss

---

7. These calculations are based on the square juger of 180' x 180' which Palladius described in 2.12. Spacing the plants at 3' would give 60 x 60 plants, i.e. 3,600 per juger as he says. For the 2½' spacing he makes a slight error: the result should be 72 x 72 = 5,184, but he gives 5,476 (= 74 x 74).

will be involved if ripe fruit is gathered together with late-ripening, since it is unsound to carry out the vintage of one variety seasonably if another unripe variety is mixed in with it, but wasteful to await the late ripening of the other variety.) §13 That is one advantage: another is that because the vintage comes in stages by virtue of the differences in the vines, a smaller number of workers is able to process it and store it variety by variety, and to succeed in keeping each wine pure-flavoured, not compromised by another variety. If this method seems difficult, you should only plant vines together that are a match both in flavour and in times of flowering and ripening.

Other planting methods

The above will be the method used in dug-over ground or trenches; in holes, however, you plant cuttings at the four corners. §14 According to Columella,[8] at the same time you will sprinkle on marc mixed with manure, and if the soil is thin you will place rich earth in the hole, even if it must be brought in from elsewhere.

When we are setting out a rooted cutting or mallet-shoot,[9] we shall place it in soil that is moderately moist (but better dry than muddy), leaving two buds above ground and setting the cuttings at an angle; in this way they will take more easily.

10. *Vines supported by trees*

Growing and transplanting the vines

§1 But if your pleasure is to have a treed vineyard, you will first need to raise plants of a superior vine in a nursery, so that once rooted they can each be transplanted from there to a hole next to a tree. By a 'nursery' we mean a bed dug over uniformly to a depth of two and a half feet. In this bed – which you can enlarge or reduce according to the number of vines (or plants of whatever kind) to be planted – you will set the cuttings at a very short distance from each other. §2 If it is valley-land or moist level ground, they can have just three

8. Columella 3.15.5, who says the recommendation comes from the agronomist Mago.
9. This is a shoot growing from a cane of the previous year, with a short stretch of the cane left attached on either side of the joint, creating a resemblance to a mallet.

buds,[10] apart from the tiny ones at the base. You will transplant the vines or saplings, now rooted, from here after two years, when they have grown strong. When you place them in a hole, you reduce each one to a single stem, pruning off all roughnesses, and also shortening any roots that you find injured. §3 Mind, in making a treed vineyard you set two rooted vines in a hole, taking care that their roots do not touch. Between them you will place stones weighing about five *librae* each, and have the vines themselves contact the opposite sides of the hole. Mago states that the hole should not be completely filled the first year, but gradually levelled up, a procedure that will make the vine sink its roots deeper. This may perhaps suit dry regions;[11] in moist ones, however, the plantings will rot as moisture enters the hole, if the earth is not mounded up straightway.

Support trees

§4 Anyone intending to make a treed vineyard should put in saplings of the following varieties, or, if they are available in abundance on the land, make use of them: poplar, elm, and ash in montane or rough places where elm is less happy. Columella says these trees too should be raised in a nursery bed.[12] My view is that since each and every one of them is produced spontaneously by all regions, sizeable seedlings or rooted trunks[13] of these species should be transplanted at this time and established beside the hole for the vine. §5 If it is grainland where you are laying out the vineyard, leave 40 feet between the trees to allow for sowing; on thin land, leave 20. A vine in its hole will need to be at a distance of one and a half feet from its tree, for a vine directly up against a tree will be smothered by the tree's growth. It should also be enclosed with a fence against damage by greedy livestock, and should be fastened straightway to its tree.

10. These would be quite short cuttings, about 7" to 8" long according to Columella (3.19.2).
11. Mago was from Carthage in North Africa. The reference to him comes from a passage in Palladius' source Columella (5.5.4).
12. Columella 5.6.5 recommends nurseries for elm and ash (but not poplar).
13. i.e. saplings reduced to trunks by removal of their leaders and branches, to reduce the stress of transplantation (cf. 3.18.1).

Propagating treed vines: a layering method

§6 There is also another method, a shortcut, for transplanting a vine from a treed vineyard. A basket is fashioned of wicker, measuring a foot or somewhat less internally from rim to rim. This is carried to the tree supporting the vine in question, and punctured at the centre of the base to admit a growing shoot of the vine. §7 So once a shoot of the vine from which you plan to make a transplant is inserted in the basket, it is hung from somewhere on the tree and filled with natural soil, so that the soil encloses the shoot. The shoot will first be twisted around. In this way, by the time a year has passed, the enclosed shoot will make roots within the aforementioned basket. Then the rooted shoot will be severed below the base of the basket, and carried (basket and all) to the place which you intend to fill with treed vines, and there it is dug in beside the roots of the tree designated as its mate. In this way you will transplant as many vines as you want without any uncertainty about their taking.

11. *Provincial vineyards*

Vineyards of many types are found in the provinces. The best type is when the vine stands like a small tree, supported on a short stock. It is aided by a cane at first, until it grows firm; it should not be taller than a foot and a half. Once mature, it will stand by itself. Another type is that in which several canes are placed around the vine and its branches are tied to the canes, so it is trained in a series of circles. The worst arrangement is when a vine stretches out over the ground and rests on it. All of these are planted either in holes or trenches.

12. *Pruning ordinary vines*

§1 This month is the right time for pruning vines in places that are temperate or somewhat cold. But where there are many vines, the pruning should be divided: those facing north should be pruned in spring, while the others, with milder exposures, should be cut back in autumn.

In pruning we should always strive to make the stock of the vine stronger, and not keep two hardened branches while the vine is young and weak. §2 We should remove shoots that are far-extended, twisted, or growing in the wrong places. Also a fork-shoot, growing centrally between two arms of a vine, should be rubbed off. But if its luxuriant growth has weakened a nearby arm, the latter should be cut off and the fork-shoot allowed to take its place. A first-rate pruner will always retain a lower shoot, growing in a good place, with a view to restoring the vine, and leave this shoot with one or two buds.

§3 In milder locations it will be possible to let the vine grow higher, but in thin-soiled or sweltering or sloping or storm-prone places it should be kept lower. In places with rich soil, allow two whips to each of the vines' arms. But a prudent person will assess the strength of the vine. One that is trained higher and is fertile should not have more than eight branches, always bearing in mind the guard-branch at the lower level.[14]

§4 Any growth around the trunk should be cut off, unless the vine needs to be restored to vigour. If the trunk of the vine has been hollowed by sun or rains or harmful animals, we clean out the dead material and smear the wounds with *amurca* and soil, a treatment that will help protect against the said troubles. Also bark that is torn and hanging from the vine should be removed, which will result in less dregs in the wine. Moss, wherever found, should be scraped off. §5 Pruning-cuts on the vine's hard wood should be sloping and round. While you cut off, as I said above, everything that is old or growing poorly, keep the new fruit-bearing shoots. Also cut back dry year-old 'claws' on the guard-branches, and any old or scaly material that you find.

Those vines that are trained higher, as on a bar or pergola, should have four arms when they stand four feet above ground. §6 If the vine is thin, we leave one whip per arm, but two if it is lush. But you must ensure that the shoots you keep are not all on one side: when that happens, the vine dries up as if touched by lightning. Shoots should not be left either around the hard wood or at the ends, since the former, like stock-shoots, are less productive, while the latter burden the vine with excessive produce and extend its growth too far. Accordingly those we preserve should be in the middle section. A cut should not be made next to a bud but somewhat higher, and it should face away from the bud because of the trickle of sap.

14. The guard-branch is the shoot mentioned at the end of §2, kept in reserve to replace one of the older branches.

13. *Pruning a treed vineyard*

§1 A vine that is placed on a tree should be cut back on its first firm-wood stem to the second or third bud. Every year thereafter we should gradually allow some growth along the tree's branches, while always directing one stem towards the top of the tree. Those who want the maximum amount of fruit allow many stems to grow along the branches, while those who want better wine direct the shoots towards the treetop. The stronger tree-branches should have more stems placed on them, and the weaker ones fewer.

§2 The system of pruning is that the old shoots, on which the previous year's fruit hung, should all be cut off and the new ones left, with their tendrils and useless little branches cleared away. We should ensure that each year the vine is loosened and refastened,[15] since that refreshes it.

We must shape the branches of the vine-bearing trees in such a way that one is not aligned with another above it. In a rich locale the elm should be left without branches for eight feet from the ground, and in thin soil for seven feet. §3 Where the ground is subject to dew and mist, the branches of a vine-bearing tree should be directed to the east and west by pruning, so that the flanks are open and can expose the entire vine's limbs to the sun's rays. We must make sure that the vine is not crowded in the tree. As the first trees weaken, they need to be replaced by others. In a sloping locale the trees' branches should be kept lower, but higher in poorly drained level ground.

The vine shoots should not be fastened to the tree with hard withies, lest the fastening cut or chafe them. You need to know this because any growth on the shoot that hangs outside the fastening will be covered with fruit, while anything inside the fastening is destined to be a firm-wood stem for the following year.

14. *Pruning provincial vines*

If you want to establish the provincial-style vines that I mentioned as standing like small trees (3.11), you will leave them branches on their four sides, and on these arms you will keep shoots in accordance

---

15. i.e. in a different place.

95

with the vine's capacity. But those vines that are trained in a circle on canes should be pruned in the same way as those supported by vine-stakes or posts. As for those that lie prostrate without supports (which should only be done out of need or the constraints of the province),[16] in the first year they will have two buds, and later more. A vine of this type needs to be pruned more closely.

15. *Pruning a young vine*

§1 Columella says a young vine should be shaped from its first year to a single firm-wood stem, and should not be cut back completely at the end of its second year (as the practice is in Italy);[17] his reason is that vines either die when completely cut back, or produce infertile shoots, which are forced to emerge from hard wood like stock-shoots since the head of the vine has been sheared off. §2 Consequently, he says, one or two buds should be left near the point where an old shoot joined the branch (something which should certainly be practiced in the case of a stronger vine); and naturally the young vine should be supported on canes or little posts, before adapting to stronger ones in its third year. §3 A young vine in its fourth year will rightly be made to sustain three firm-wood stems where the soil is fertile.

Immediately after the pruning all the shoots cut from the vines should be removed, plus brambles and anything that might impede the digger.[18]

16. *Propagation by layers*

§1 This month too, vines should be layered. According to Columella, it will be better to renew an old decayed vine, whose hard growth has spread a long way, by means of 'divers', rather than bury it by digging-in its entire framework – an opinion that is sure to displease farmers. (We call it a 'diver' when an arch, so to speak, is left above ground while another section of the vine is dug in.) §2 Columella

16. The constraints are primarily those of harsh climate, which makes low growth essential as a protection against winds and cold (Columella 5.5.17–19).
17. Columella 4.11.1–4.
18. Pruning was followed by digging, to control weeds and allow penetration of air and rain (3.20.1, 4.7.2, 5.2.3).

explains that when the whole vine is laid flat, it is exhausted by the large number of roots produced all over its body. The 'divers' are severed in the above-ground section after two years, the result being proper vines *in situ*.[19] But farmers maintain that if you sever them after two years, they generally have weak roots and die suddenly en masse.[20]

Trees and Other Plants

17. *Grafting methods*

§1 This month is the best time to carry out grafting in hot sunny places. There are three methods: two of these can be done now, while the third is kept for summer. The methods of grafting are: under the bark, in the trunk, or with a 'shield'.

We shall graft, then, as follows. We use a saw to cut through the tree, or a branch, at a spot that is shiny and unscarred, without damaging the bark. After sawing we trim the wound with sharp knives. §2 Then we insert a kind of thin wedge of iron or bone (preferably from a lion) for about three fingers between the bark and the wood – circumspectly, to avoid splitting the surface of the bark. In the same way, after withdrawing the wedge, we straightway push in a scion; this has been whittled for half its length, while the pith and bark remain on the other half. It should stand out six to eight fingers from the tree. §3 We install two or three or more scions according to the nature of the trunk, leaving a distance of four fingers or more between them. Then we bind the trunk with rushes or elm[21] or osiers, put clay over it covered with moss, and wrap it, allowing the scion to project four fingers from the clay.

Many people like the following method: once the tree is cut through, they first bind the trunk quite tightly, then split it in the

---

19. Columella 4.2.2, 4.22.2–3, 4.15.2–3. Columella actually says Julius Atticus is his authority for preferring 'divers'. The laying-flat method is described at ps.-Columella *Trees* 7.4.

20. Elsewhere, perhaps because of this problem, Palladius recommends detaching divers after three years (12.2).

21. i.e. elm bark.

middle; next they take scions that have been pared on each side into a wedge shape without removing the pith, and push these in; before doing so, they insert the little wedge, and as it is withdrawn the scion is pushed in, and is held tightly by the wood as it moves back onto the wound.

§4 Both of these methods are carried out in spring and on a waxing moon, as the trees' buds begin to swell. The scions that are to be inserted should be young, fertile, knotty, growing from new wood, cut from the east side of the tree, the thickness of your little finger, with two or three forks, and well provided with several buds. §5 If you want to graft a smaller tree (which definitely produces better growth), cut it close to the ground; and it is better to insert the scions between the wood and the bark, then bind them. Some people take a scion, pared on each side, that matches the girth of the tree to be grafted, and insert it in the centre in such a way that the bark of the scion is flush with the tree's bark all round. In the case of a young tree the earth should be loosened and heaped up right to the point of the graft. This will protect it from wind and heat. §6 One meticulous farmer stated to me that any graft takes reliably if after inserting the scions we press undiluted birdlime with them into the wound, using it as a kind of glue to blend the sap of the two materials.

§7 We shall talk about shield-budding in the proper month (7.5.2–4). Columella described a fourth method as follows.[22] Using a Gallic auger, one should bore a hole as far as the pithy centre of the tree, sloping slightly down towards the inside. When all the sawdust has been extracted, a vine-stem or tree-branch, pared of its bark so as to fit the hole but full of sap and moisture, is inserted firmly, leaving one or two buds outside. Then the place is carefully covered with clay and moss. In this way, he says, it is even possible to graft vines that are attached to elms.[23]

22. Columella 4.29.13–16. Columella claims to have invented the Gallic auger specifically for grafting. Its cutting edge produced shavings (rather than sawdust like the old auger), leaving the hole clean and unscorched, and so improving the chances that the graft would take.
23. i.e. in a treed vineyard. Because they were entangled with their support trees, such vines could not be cut down, as required for grafting under the bark of the trunk.

A graft to produce stoneless peaches

§8 A person from Spain showed me the following new grafting method, assuring me that he had tested its effectiveness on peaches. His instructions were that a willow branch the thickness of your arm, with no cavities, two cubits or more in length, should have a hole bored through the centre. Next a young peach plant, without being moved from its location, is stripped of all its branches, with only the leader left, and is guided through the hole in the willow shaft. Then the willow branch should be curved into an arch shape, with each end sunk in the ground, and the hole should be sealed with mud, moss and bindings. A year later, when the young plant's leader has adhered so closely within the willow pith that a single whole has formed from the two bodies, the plant is severed beneath and transplanted, and sufficient earth is heaped up to cover the willow arch along with the peach leader. As a result, he reported, the peach fruits grow without stones. But he added that this method is suited to moist or irrigated areas, and that the willows should be helped with watering, so that the moisture-loving nature of the wood can flourish, and can dispense its superabundance of sap to the foreign growth.

18. *Establishing olive orchards*

§1 This month in temperate areas we shall establish olive orchards. They should either be planted in dug-over ground round the edges of the vine beds on each side of the walkway,[24] or else they will have their own special area.

If they are placed in dug-over ground, rooted saplings with their leaders cut off and their arms shortened to one cubit and a palm from the trunk should be planted in the aerated dug soil, with a post used to sink the holes. Barley grains should be thrown in beneath them, and they should be docked of any matter found to be decayed or dried out; then their tops should be covered in mud and moss, and wrapped with fastenings of elm bark or bindings of any kind. §2 But the greatest benefit and asset to their growth is if the direction which they stood facing is marked on them with

24. i.e. in vineyards laid out with separate beds for different varieties, as prescribed above (3.9.11).

ruddle, and they are planted similarly looking in that direction. They should be at a distance of 15 or 20 feet from each other. All herbage around them should be pulled out regularly, and whenever there are rain showers, the trees should be stimulated by very shallow but very frequent cultivation of the ground. Also the soil should be drawn away from the trunk now and then, mixed up and piled somewhat higher.

§3 But if you want to make an orchard for olives only, you will look for the following types of soil: soil that has gravel mixed with it, or clay combined with grit and broken up by it, or rich grit, or soil of a denser fertile nature. §4 Potter's clay must be rejected absolutely, and boggy soil, and that which always has standing water, and lean grit and bare gravel; for although the trees may take, they do not flourish. They can be planted where arbutus or holm-oak stood previously, but not Turkey oak, for even when cut down it leaves harmful roots, whose poison kills the olive trees. In sweltering places they enjoy a north-facing hill, but a south-facing one in cold places, and in middling climates they like slopes; and they do not tolerate a low-lying situation or a high one, but prefer moderate slopes as in the Sabine region or Baetica.[25]

The types of olive berries are numerous and have many names, such as pausian, orchis, radius, Sergian, Licinian, Cominian and all the others that need not be named here. The oil yielded by the pausian is excellent while fresh, but soon spoils with age. The Licinian gives the best oil, the Sergian the largest quantity. It will suffice to give the following general advice about these varieties: larger berries will be useful for food, smaller ones for oil.

§5 If it is grain-land that we are planting as an olive orchard, they should be 40 feet apart,[26] but 25 feet if the land is thin. We shall do better if we align the rows towards the west. When they are planted, they should be set in dry holes four feet deep. Also gravel should be mixed in where there is a lack of stones, and manure. If the place is enclosed, the trees as they are planted should project only slightly above ground level; but if livestock are a threat, the trunks will need

25. The Sabine region extended to the north-east of Rome, including the slopes of the Apennines. Baetica is south-central Spain (roughly modern Andalusia).
26. Grain crops will then be grown between the rows of olive trees.

to be higher. In dry regions it will also be appropriate to irrigate when there is a lack of rain.

§6 If the region lacks olive orchards and there is nowhere from which young plants can be taken, you must make a nursery, i.e. a dug-over bed, as I said above (3.10.1–2). There, as Columella says,[27] branches sawn into one-and-a-half-foot lengths can be planted. After five years the plants will be strong and capable of being transplanted from there; they can be planted this month in cold places. I know most people do something easier and more practical: they find the olive trees that grow widely in woodlands or uninhabited places, cut their roots into cubit lengths, set them out either in a nursery if desired or in an olive orchard, and assist them by mixing in manure. The result will be that numerous plants grow from one tree's roots.[28]

19. *Planting fruit trees*

§1 We can also set out fruit trees in trenched ground in north-facing areas. We shall discuss their special requirements individually. The kind of ground that suits vines also suits fruit trees, but you will make the holes bigger, to help the growth of wood and fruit. If you are making an orchard, you will leave 30 feet between the rows. §2 Installing rooted plants is the better way.[29] Take care that their tops are not broken by handling, or browsed, leaving them unable to grow. You will assign each row to a particular species, to avoid weak trees being overwhelmed by stronger ones. We shall mark the plants similarly,[30] so we can plant them in the orientation in which they were standing. §3 We shall transplant from sloping, dry, thin locations to level, rich moist ones. If you want to plant trunks,[31] they should project about three feet above ground. Where you are setting two plants in one hole, take care that they are not touching, or they will be killed by worms. But as Columella says, trees grown from seed, i.e. from their own nuts, are more productive than those

27. Columella 5.9.3.
28. Cultivated olives would then be grafted onto these wild olive stocks (cf. 5.2.1–2).
29. Better than unrooted branches, which would be slower to establish themselves.
30. i.e. as with olive plants, 3.18.2.
31. i.e. 'rooted trunks' in the sense used at 3.10.4.

planted from slips or from lengths of branch.[32] Where the region is drier, they should be helped by watering.

20. *Maintenance of vines and trees*

§1 In seaside or hot places the vineyards should be dug now, or ploughed if that is the custom of the region.[33] In the same places the vines should be staked and fastened before the bud swells, since great losses are incurred if one knocks or chafes the bud.

§2 Now olives and other trees should be fertilized on a waning moon. One cart-load will be enough for a larger tree, and half for a smaller one; the earth should be dug away from the roots, mixed with the dung and put back.

At this time one should dig around any young plants in the nursery beds, and prune off superfluous branches and any higher rootlets growing round them.

21. *Flower beds*

§1 This month we shall start rose beds. They should be planted in a very shallow trench or in holes, using either cuttings or even seed. We should not think rose seeds are the golden centres of the flowers carried by roses: rather they develop berries which ripen, looking like very small pears and full of seeds, after the grape harvest. Their ripeness can be judged by their dark colour and softness. §2 Old rose beds are dug around at this time using hoes or picks, and all dryness is cut away. Now too beds that have gaps can be restored by layering branches. If you want to have roses earlier, you will dig a circle two palms' breadth from the stock, and water it with hot water twice daily.

§3 Now too we shall plant lily bulbs, or hoe lilies that were established previously – with the greatest care, to avoid wounding the 'eyes' that grow around the root, or the smaller bulbs. These, when

32. This may be a mistaken memory, by Palladius or an intermediary, of Columella's statement (5.10.6) that a grafted tree is more fruitful than those planted by cut branches or slips.
33. This cultivation between the rows of vines controlled the weeds, and allowed water and air to reach the roots of the vines.

pulled off the mother bulb and set out in rows elsewhere, will form new lily beds.

Again, we should plant out violet sets and saffron bulbs, or delicately dig around those already in place.

22. *Sowing flax*

This month some people sow flax seed in fertile soil at 10 *modii* to the juger, and obtain slender flax threads as a result.[34]

23. *Utilitarian plants*

§1 At this time beds of *reeds*[35] should be planted, using very shallow holes and burying the 'eyes' of the reeds each in its own hole, half a foot apart. If we are dealing with a hot dry region, we need to earmark for the reed beds valley land that is moist or irrigated; if the area is cold, they should be established in places of middling elevation that receive the outflow from farm buildings. Among them we can also scatter *asparagus* seeds, so they can grow as a companion crop, since asparagus is cultivated and burnt off in the same way as reeds. §2 But if one has old reed beds, one will hoe them at this time, cutting away anything on the root that needs cleaning, i.e. anything decayed, crooked, or lacking 'eyes' for growth.

Now we shall put in young plants of *willow*, and of all the species utilized for treed vineyards;[36] also *broom* where it is not present.

We shall also make nursery beds using *myrtle* and *laurel* berries, or cultivate existing beds.

24. Gardens

§1 Around the middle of February garden hedges should be created, from thorn seeds embedded in ropes, as was said when we were talking about protecting gardens (1.34.5). Alternatively, the Greeks say one

34. The dense sowing recommended here would result in crowded plants and consequently in fine fibres, desirable for weaving linen. Another method of producing thin-stemmed plants is to sow the flax thickly in poor soil (11.2).
35. The reference is to the giant reed or giant cane, *Arundo donax,* which was used for stakes and in construction.
36. Chiefly poplar, elm and ash (3.10.4).

should cut small sections from a thick bramble stem, bury them in palm-deep holes, and tend them daily by digging and watering until they leaf out.

§2 *Lettuce* is sown this month, for planting out in April. *Cardoon* too is sown, also *cress, coriander* and *poppy*; *garlic* too, or *ulpicum*, as in November.

Now *savory* is sown, in rich ground, manured, as long as it is sunny, or (better still) near the sea. It can also be sown in combination with onions.

§3 Also this month you sow *onions* – though it is generally agreed that both spring and autumn are good sowing times. If you sow the seed, it grows into a bulb and yields less in the way of seed; if you plant the bulb, it decreases in size but produces much seed. Onions want rich soil, vigorously worked over, watered and manured. We shall make beds in such soil, cleaned of all weeds and root. We shall sow on a calm fine day, preferably when the wind is from the south or east. §4 If sown on a waning moon, their growth is skinny and rather bitter; if on a waxing moon, they are firm and succulent in flavour. They should be planted fairly thinly, and weeded and hoed quite frequently. If we want bigger bulbs, we should remove all their leaves: in this way the juice will be forced to the lower end. Those for seed-collection should be supported with small stakes when they begin to develop a stalk. The seeds signal their ripeness when they turn black in colour. The stems with their seeds should be picked while still half-dry, and dried like that in the sun.

§5 This month in cold places you will sow *dill*. It stands any climate but enjoys a warmer one. Water it if rain holds off; sow it fairly thinly. Some people do not bury its seed, opining that no birds touch it.

Now too we can sow *white mustard*.

This month too we shall sow *cabbages*, though they can actually be sown throughout the year. They love rich well-worked soil, fear argil and gravel, are not happy with grit or sand unless helped by a year-round water supply. §6 Cabbage stands any climate, preferably a cold one. They produce sooner if facing south, later if facing north, but the latter win out in flavour and sturdiness. They like slopes, and so should be planted out on ridges in the beds. They enjoy dung and hoeing. If planted fairly thinly they grow strong. They will cook quicker while keeping their crisp greenness, if you sprinkle on ground *nitrum*[37] through a sieve, so as to look like hoarfrost, when they have three or four leaves. §7 Columella says that to help them keep their fresh greenness,[38] we should wrap their roots in seaweed when transplanting, and daub on some dung at the same time. One should plant the seedlings that have more growth, because, although they take more slowly, they will become stronger. In winter, transplanting should be done in the warmth of the day; in summer, as the sun dips towards evening. They will grow bulkier, if regularly covered with soil. Old cabbage seed turns into turnip.

§8 After the middle of this month we shall start *asparagus* 'sponges',[39] either growing new ones from seed or planting out old ones. Actually I think it is a cheap and thrifty method if we plant plenty of wild asparagus roots in a spot that is uncultivated (or certainly stony), so they can yield a harvest straightway from a place that would not support anything else. And we shall burn these each year, i.e. the stalks, so their yield can come up denser and stronger. This variety is more pleasant in flavour.

§9 Now too *mallow* can be started.

Start *mint*, too, with seedlings or roots, in a moist place or around water. It wants sunny ground, neither rich nor manured.

This month you will start *fennel* in a sunny place, moderately rocky.

Carrots[40] are started in early spring. They are planted by seed or seedlings in rich ground, loose, quite deeply trenched. You will set them out well spaced, so they can grow strong.

Summer savory is sown now as well, and cultivated in the same way as garlic or onion.

Now is the time for sowing *chervil* in cold places, after mid-month; it wants land that is fertile, moist, manured.

§10 This month we shall sow *beets*, though they can be sown throughout the summer. They love crumbly land and a moist location.

37. A term covering various alkalis, especially potash and soda.
38. i.e. in cooking: Columella 11.3.23.
39. Described at 4.9.11.
40. The Latin *pastinaca* actually designates both carrot and parsnip: these plants may have been more similar in their cultivated forms than they are today.

They are ready for transplanting (roots smeared with fresh dung) when four- or five-leaved. They love being dug around frequently and drenched with lots of manure.

§11 *Leeks* are to be sown this month. If you want them for cutting, you can cut them two months after sowing, leaving them in the beds, though Columella states that even cutting leeks will last longer and better if transplanted.[41] Each time they are cut, they should be helped with water and manure. If you want to make them 'headed',[42] you will need to transplant in October what you sowed in spring. They should be sown in fertile ground, preferably level, in a bed that is flat, deeply trenched, long worked over, and manured. If you want cutting leeks, you sow more thickly; if 'headed', more thinly. They need to be frequently visited with the hoe and kept free of weeds. §12 When they are the thickness of a finger, they should be transplanted, with the leaves chopped off halfway up, and the roots cropped; smear them with liquid dung, and set them four or five fingers apart. Once they make roots, they should be grasped circumspectly and raised with the hoe, so they are held away from the soil, and consequently forced to fill the empty space that they find beneath them with a head of great size. There again, if you plant several seeds wrapped together, a large leek will grow from all of them. Likewise, if you insert turnip seed in the head (without using a knife) and plant it, people say it puts on much growth – all the better, if you do this frequently.

§13 This month *elecampane* is started where reed beds are planted. It is started by eyes like reeds; we need to cut them off and bury them lightly with earth, in trenched well-worked soil, after raising linear ridges into which we should dig the eyes. They are separated from each other by a three-foot space.

§14 This month we shall plant the bulbs of *Colocasia*.[43] They love a moist rich place, especially one that is irrigated. They are happy around springs and streams, and do not care about soil quality if fostered by constant moisture. They can put out leaf almost continuously, if they are protected by covers from the cold like citron-plantings.

41. Columella 11.3.30.
42. The 'head' is the bulbous base.
43. Probably a variety of taro, *Colocasia esculenta*; the 'bulbs' would be the corms.

This month *cumin* and *anise* are sown, in a well-worked place into which you should mix fertilizer. The sowings should be regularly cleared of weeds.

25. Fruit Trees

Pears: propagation

§1 We shall plant pear seedlings in February in cold places, but in November in hot ones. The reason for starting pears in November in warm places is that they can benefit from the moisture in the soil; in this way they will not only flower abundantly, but also swell the fruit to a good size. They particularly love to grow in the kind of soil that we noted as suitable for vines (1.5.4, 2.13.5), but in fertile soil we shall produce strong trees and plenty of fruit. §2 It is believed that pears of a stony type will lose this fault, if started in soft ground.

Actually to start pears from seedlings virtually guarantees slow results. However, those who choose this method, so as to keep the cultivated stock unalloyed by the harsh wild varieties, should plant them like olives, placing them in big holes when two or three years old and well rooted, so as to project three or four feet above ground. Moss mixed with clay should be used to cover the severed tops.[44] For if a person scatters pear seeds, they will undoubtedly grow: nature always fosters her own method of reproduction, and to her eternal perspective no amount of slowness can be wearisome. But for a human being it is a protracted business to wait for this, since they both develop slowly and degenerate from the good breeding of the type. §3 So the better method will be if in November we start rooted seedlings of wild pears in holes in well-worked ground, to be grafted once they have taken. There is this difference, that those started from pure-bred seedlings preserve their sweetness and tenderness, but do not keep long in storage, whereas those that are grafted will stand the passage of time.

A measurement of 30 feet should determine the distance between pear trees. To produce results this species needs to be nurtured

44. Their leaders have been removed, as was recommended above when transplanting young rooted olives (3.18.1).

with frequent watering and repeated digging – so much so that in flowering season it is thought not to lose any of the flowers it has produced, if the digger helps it at that time.[45] §4 You will do much good if after a year's interval you supplement it with manure: any kind will do, but ox manure is thought to produce close-set heavy fruit. Some people mix in ash, believing that this concentrates strong flavours in the fruit.

I think it superfluous to set out the various types, since there is no difference in planting and cultivation.

Pears: ailments

If a pear tree is sickly, you either ablaqueate, bore a hole in the root and force in a wooden stake, or else bore a hole similarly in the trunk and fix in it a wedge of pine or, failing that, of oak. §5 The worms of this tree are prevented from emerging, or killed if they emerge, by infusing the roots frequently with bull's bile. Again, fresh dregs of old wine, if poured on the roots for three days, prevent the trees from labouring too long in flower. §6 If a pear is stony, you will lift the existing earth from the tips of the roots and separate out all the little stones. After carefully removing them you pour in its place fresh earth which has been sieved. Mind, this will only be effective if you are consistent with watering.

Pears: grafting

Pear is grafted in February and March, in the manner described when we were speaking of grafting (3.17), under the bark or in the trunk. It is grafted on wild pear and apple; according to several authorities, on almond and thorn; according to Vergil,[46] on flowering ash; also on ash and quince; according to some, on pomegranate too (but only in the split trunk). §7 A pear scion inserted before the solstice should be a year old, and before insertion it should be stripped of its leaves and all soft wood; but after the solstice you insert one that includes a terminal shoot.

---

45. For other plants the rule is not to cultivate at flowering time, in order to avoid knocking off the blossom and so reducing the crop (e.g. 2.15.9).
46. *Georgics* 2.71–72.

Pears: preserving

§8 Pears should be preserved on a calm day, on a waning moon, from the twenty-second to the twenty-eighth day of the moon. The fruits should be dry, and picked by hand from the second hour to the fifth or from the seventh to the tenth; they should be sound and almost hard and somewhat green, carefully differentiated from those ready to fall. They are stored in a pitched vase, which is then covered with a lid, tipped mouth-down and buried in a place near which there is perennial water running. §9 Then again, pears that are hard in flesh and skin are first placed in a heap where they begin to soften, and then placed in an earthenware vessel (well fired and pitched) and sealed with plaster round the lid; the vessel is sunk in a shallow hole in a spot reached by the sun every day. Many people have kept pears buried in chaff or grain. Others have stored them in pitched jars immediately after picking them with their stalks, then closed the mouths of the vessels with plaster or pitch, and buried them in the open under a covering of grit. Others have kept pears, not touching each other, in honey. Then again, pears are cut open, cleaned of their seeds and dried in the sun. §10 Some heat salt water and skim off the foam when it begins to boil; after it has cooled they drop in the pears to be preserved, then take them out after a short time, cache them in a jar, plaster its mouth and store them; or they let them stay a night and a day in the cold salt water, next steep them for two days in fresh water, then preserve them plunged in *sapa* or raisin wine or sweet wine.

Pear wine and vinegar

§11 Wine is made from pears by bruising them, enclosing them in a very fine bag, and pressing them with weights or in a press. It lasts the winter, but sours at the start of summer.

Vinegar is made from pears as follows. Wild pears, or those of a harsh variety, are stored when ripe in a pile for three days. Then they are dropped in a vessel with the addition of spring water or rain water, covered and left for 30 days; thereafter the vinegar drawn off for use is replaced with an equal quantity of water.

Pear sauce

§12 Ceremonial pear sauce [47] can be made as follows. Very ripe pears are trodden with whole salt. When their flesh has been pulped, they are stored in small casks or in pitched earthenware vessels. When drained after the third month, this flesh releases a liquid of delightful flavour but pallid colour. To correct this it will be helpful if you mix in a certain amount of darkish wine when they are being salted.

Apple trees

§13 We shall start apple trees in February or March – but in October or November if the area is hot and dry. There are many varieties, which it would be superfluous to enumerate. They love rich fertile soil whose moisture is supplied not so much by watering as by nature. (But a tree in sand or argil should be helped by watering.) In montane places they should be established facing south. They grow in cold soil too, if helped by the warmth of the atmosphere, and they do not refuse a position in rough or moist ground. Lean dry soil makes the apples wormy and liable to fall.

§14 They are started by all methods, just like pears. They do not like to be ploughed or dug around, and for that reason meadows suit them better. They do not absolutely require sheep dung, but gladly accept it, at least if fine ash is mixed with it. They like moderate watering. Pruning is appropriate, but chiefly to remove dry or malformed wood. §15 This tree ages quite quickly, and deteriorates in old age.

Apple trees: remedies

If the apples tend to fall, a stone inserted in a root that you have split will hold them in place. If the treetop is touched with the gall of a green lizard, the tree does not rot. Its worms are dispatched by pig manure mixed with human urine, or by ox gall. If there are large numbers of them around a tree, once scraped away with a bronze paring-knife they will not appear any more. The scraped areas should be covered with ox manure. §16 If the branches are burdened by a heavy crop of apples, all the flawed fruit should be thinned out, so

47. Possibly such sauce was originally made for cult ceremonies in which alcohol was forbidden; however, note Palladius' suggestion of adding dark wine for colour.

the sap can share nourishment equally among the rest, and provide the well-bred apples with the abundance that was being wasted on a multitude of mean ones.

Apple trees: grafting

§17 Apple can be grafted by all the same methods as pear. In February, March and other months as for pear it is grafted on apple, pear, thorn, plum, service tree, peach, plane, poplar, willow.

Apples: preserving

Apples that we want to store should be gathered carefully. We lay them out grouped in piles in dark places where there is no wind, after first putting down straw on hurdles. The piles should be separated by frequent divisions. Some people mention diverse methods: either enclosing them individually in pitched sealed earthenware vessels, or coating them with argil, or just daubing their stalks with clay, or laying them out on boards spread with chaff and covering them with straw on top. §18 Round apples, called 'globulars', can keep for a whole year without any care. Some people sink earthenware vessels in a well or cistern, entrusting apples to them after carefully pitching and closing them. Some have taken undamaged apples from the tree, plunged their stalks in boiling pitch and laid them out in rows on a wooden floor with walnut leaves spread beneath them. Most people pour poplar or fir sawdust between the apples. §19 Everyone agrees that you should place the apples stalk down, and not touch them before you feel the need to use them.

Wine and vinegar are made from apples, in the way that I instructed earlier from pears (§11).

Quince

§20 Different dates for starting quinces are mentioned by many authors. But I discovered from experience that in Italy around the city, in February or the beginning of March, rooted quince plants took so successfully in trenched soil that they often gloried in fruit the following year, if they had been of considerable height when planted. In hot dry places they should be planted in October or the beginning of November. Quinces love a cold moist place; if they are established in a warm one, they need to be helped by watering. §21 But they accept placement on a middling site, between hot and cold in nature. They grow in both level situations and shelving ones, but prefer slopes and inclines. Some start them from branch-tips or cuttings, but progress is slow by either method. Quince trees should be placed in such a way that one is not touched by drips from another when the wind shakes them.

§22 While it is smaller or being planted, it should be helped with dung, but when bigger with ash or powdered chalk sprinkled on the roots once per year. Constant moisture will make the fruit of this species early-ripening and larger in growth. They should be watered whenever heaven's showers are denied, and dug around in October or November in hot places, but in February or March in cold ones. If it is not dug around regularly, the tree is either rendered sterile or else its fruit degenerates. It does need to be pruned, as I have found, and freed of everything defective.

Quince: remedies
§23 If the tree is sickly, *amurca* mixed in equal parts with water should be poured on the roots, or quicklime combined with chalk or else box-resin [48] mixed with liquid pitch should be smeared on the tree trunk; others state that the tree should be ablaqueated and an uneven number of quinces, commensurate with its size, should be placed round its roots and buried. This, done every year, will protect from disorders, but it will subtract from the tree's long life.

Quince: grafting
§24 Quinces are grafted in February, in the trunk rather than the bark. They accept being grafted with scions of almost every type, pomegranate, service tree, and all the soft-skinned fruits (which they improve). Grafting should be done on young trees, which have sap in the bark. A larger tree is better grafted near the root, where the bark and wood are moist thanks to the soil clinging to them.

Quince: preserving
§25 Quinces should be picked when ripe, and saved as follows:

48. Apparently a type of resin that was stored in boxes.

either placed between two tiles, if enclosed with mud all round, or else stewed in *defritum* or raisin wine. Others keep the bigger ones wrapped in fig leaves. Others just put them aside in dry places, from which the wind is excluded. Others, after cutting them in quarters with a cane or ivory knife and removing all the centre, cover them with honey in an earthenware vessel. §26 Others place them whole, just as they are, in honey; fairly ripe fruit is selected for this method of preserving. Others bury them in millet or plunge them in chaff, not touching each other. Others put them in vessels full of best wine, or make an equal mixture of wine and *defritum* for keeping the quinces. Others plunge them in jars of must and then seal them, a process that makes the wine fragrant too. Others cover the quinces, not touching, with dry gypsum in a new dish.

Carob

§27 Carob is started in February or November, both by seeds and by plants. It loves places that are maritime, hot, dry, flat; however, as I have found by experience, it will become more fruitful in hot places if it is helped with moisture. It can also be planted by cuttings. It wants a fairly large hole. Some believe it can be grafted in February onto plum or almond. Carobs keep for a very long time, if they are spread out on racks.

Black mulberry

§28 The mulberry is friendly with the vine. Mulberries grow from seed, but then both the fruit and the woody growth deteriorate. It should be started from cuttings or branch-tips, but better from 18" cuttings that have been smoothed on each side and smeared with dung. After first making a site with a post, we plunge them in and blanket them with ash mixed with soil. We cover them no deeper than four fingers. We shall start them from the middle of February on and throughout March, but in hotter places at the last of October or the start of November; in spring the best date is 24th March. §29 They love hot places with gritty soil, and in general those by the sea. They take only with difficulty in tufa or argil.

Constant moisture is not believed to be helpful to mulberries. They enjoy being dug around and manured. Rotten and dry wood should be pruned in this species after the third year. You will transplant the sapling, if sturdy, in October or November, but if tender, in February or March. It wants deeper holes and wider spacing, so as not to be crowded by the next tree's shade. §30 Some have recounted that a mulberry tree becomes fruitful and more vigorous if we bore a hole in the trunk on each side and insert a wedge of terebinth on one side, and of mastic on the other. Around the end of September the mulberry should be ablaqueated, and very fresh dregs of old wine poured on its roots. It is grafted onto fig and onto itself, but only under the bark. If grafted onto elm it takes, but produces growth of great infertility.

Filberts

§31 Filberts should be planted by means of their nuts. Earth should be drawn over them to a depth of not more than two fingers. In my experience, however, they come better from slips and suckers. Either slip or seed is planted out in February. They enjoy a place that is lean, moist, cold, even with gritty soil. Filberts are ripe in July, around the 7th.

Sebesten

§32 Now sebesten is planted, by nuts placed in a container until the plants acquire some robustness, in a warm climate with loose soil and moderate moisture. They are grafted in March onto service tree or thorn.

Other fruits, discussed under other months

Now too *azaroles* are sown and grafted. Stones of hard-fruited *peaches* or slips of the same variety are planted; these peaches are also transplanted, and can be grafted.[49] *Medlar* too is grafted, and *plum* stones planted. §33 The *fig* also can be planted in temperate places, and the *service tree* can be started in February as well. *Almond* seeds can be planted in beds, and they can be grafted now – at the start of the month in temperate places, and at the end in cold ones, but the scions should be stored before they blossom.[50] *Pistachio* plants can

49. Throughout this paragraph, 'grafted' means 'used as grafts' on other trees.
50. i.e. they should be cut earlier in the winter and stored, cf. 2.15.12.

be established even now, or they can be grafted, and *chestnut* seed scattered. Also *walnuts* too can be buried this month too in nursery beds, and the species can be grafted; and in cold or moist places *pine* plantations can be seeded.

Livestock

26. *Raising pigs*

§1 Now above all is the time when the boars will need to breed the females. The boars chosen should be large and ample-bodied, but rotund rather than long, with large belly and haunches, a short snout, and a neck thick with glands; they should be libidinous yearlings, and they are capable of breeding the females till they are four-year-olds. As breeding sows we should choose those with long flanks and a big roomy belly to sustain the burden of pregnancy, and in other respects like the boars; §2 in cold regions they should have a dense black coat, but in warm regions just as it comes. With regard to reproduction, a female will be capable of bearing the burdens of gestation until year seven; with regard to conception, she should begin as a yearling. They give birth after four months, at the beginning of the fifth.

They are started, as I said, in February, so that the offspring can feed on more substantial herbage and on the stubble that follows. Where there is the opportunity of marketing, the successive batches of offspring are sold;[51] this restores the mothers to breeding capacity more quickly.

§3 This species can be kept in all kinds of locales, but it does better in marshy fields than dry ones, especially if woodland with fruit-bearing trees is available to help out at intervals with ripe fruit in the annual sequence. They feed particularly in grassy places and on the roots of reeds or rushes. But when food is short in winter they should sometimes be provided with acorn fodder, chestnuts or cheap left-overs from the other crops; more so in spring, when fresh greenstuff is sappy and tends to harm the pigs.

---

51. i.e. while still suckling, rather than going on later to solid food, as in the previous sentence.

§4 Sows should not be penned together like other livestock; rather we shall make sties under colonnades, in which each sow can be penned. Here she will be safer herself, and can protect her litter from the cold. These sties should be open on top, so the herdsman is unhampered as he checks the numbers and frequently rescues the offspring by pulling them out from under the mother when they are pinned. He will take care that the offspring penned with each mother are her own. According to Columella, she should not rear more than eight.[52] §5 But it is more practical, in my experience, that a properly fed sow should rear six at most: although she can raise more, nevertheless she will lose strength if suckled by a larger number.

In addition, the following is a useful trait in pigs: if let into a vineyard before the grapes swell or after the vintage is completed, they equal the diligence of a human digger in attacking the grass.

Miscellaneous: Wines and Vines

27. *Myrtle wine*

At the beginning of this month you will make myrtle wine by another method as follows.[53] You place 10 *sextarii* of old wine in a flagon, and mix in five *librae* of myrtle berries. When they have spent a period of 22 days in this blend – during which the vessel should be shaken daily – you will then strain it through a palm basket, and mix five *librae* of the best honey, vigorously whipped, with the aforesaid 10 *sextarii*.

28. *Antitoxin vine*

§1 We shall make an antitoxin vine as follows. Its value is that its wine or vinegar or grapes, or ash from its branches, is efficacious against the bites of all kinds of beasts. It is made as follows. The cutting to be planted should be split at the bottom for a distance of three fingers, and after the pithy centre has been removed, antitoxin medicament should be added in its place. Then it should be committed to earth, carefully bound up. §2 Others stow the same cuttings, once

52. Columella 7.9.13.
53. The first method was recounted in January (2.18).

impregnated with medicament, in a squill bulb, and commit them to earth in that fashion. Others steep the roots of the vine with an infusion of the antidote. Of course, if a cutting is taken from this vine for transplanting, it will not retain the medicinal power of the mother vine. One should renew the strength of the vine's ageing sap with regular infusions of the antitoxin.[54]

29. *Seedless grapes*

§1 There is a fine type of grape that contains no seeds. The result is that one can enjoy the flesh of a whole bunch of grapes like a single fruit, with the greatest pleasure and no hindrance. According to the Greek authorities it is done this way, with nature being improved through human skill. We shall have to split the cutting to be buried for the length that will be hidden in the ground, and after removing all the pithy centre and carefully scraping it out we shall bring the arms of the split section together again, bind them up and plant them. §2 They state that the binding should be made of papyrus, and that this bound section should be placed in moist earth. A more painstaking method used by some people is to stash the part of the cutting that has been hollowed and bound in a squill bulb; they state that all plantings take more easily with this assistance. §3 Others, when pruning the vines, hollow out a fruit-bearing shoot of the pruned vine on the vine itself, removing as much pith as they can from above without splitting the shoot, and fasten it to a cane as a splint, to prevent it from tilting down. Then they pour Cyrenaic juice,[55] as the Greeks call it, in the hollowed-out part, after letting it down with water to the thickness of *sapa*, and they renew this every eight days until the vine puts out new growth. The Greeks maintain that this can also be done with pomegranates and with cherries: this needs to be tested.[56]

---

54. Vines treated in this way were intended to provide quick access to medication for people who suffered bites or puncture wounds while working in the fields (cf. 14.1.5).

55. This is asafoetida, the smelly gum of the silphium plant which grew around Cyrene.

56. At 11.12.7 Palladius recounts a method of treating a young cherry tree so as to produce seedless fruit. His source there is Gargilius Martialis.

30. *Excessively tearful vines*

Vines that languish through an excess of tears, and stop fruiting as a result of weeping away their vigour: the Greeks advise that the trunk should be gashed and a recess formed; and that if this has little effect, the thick wood of the roots should be cut open, so the wound created can effect a cure. Then the excised opening should be smeared with unsalted *amurca* boiled down to half quantity and cooled, and sharp vinegar should be poured beneath it.

31. *Myrtle wine: another Greek recipe*

§1 Again, the Greeks instruct that myrtle wine should be blended as follows. You place eight ounces of ripe myrtle berries, dried in the shade and then pounded, in a linen cloth, hang it in the wine, cover the vessel and seal it. After it has been there many days, remove and use. §2 Some people take ripe myrtle berries gathered from fairly dry places with no rain on them, tread or press them and mix the juice with wine, measuring eight *cotulae* to an *amphora* of wine. This wine will also be useful medicinally when one needs to use astringents: they say it binds a queasy stomach, suppresses vomiting of blood, curbs flux from the belly, and solidifies therapeutically the mucus caused by dysentery.

32. *Spontaneous production of aperitifs*

It is said that spiced wine, or wine flavoured with wormwood or rose or violet, comes spontaneously from vines – with nature undertaking what is usually achieved by human work – if you plunge vine-cuttings in a vessel half-full of the above-mentioned potions, and at the same time dissolve natural soil in the potions, as in making lye. You keep the cuttings there until their 'eyes' are trying to open; then you plant the budding cuttings in whatever location you wish, just like the other vines.

33. *Producing variegated grape-clusters*

For a vine to bear grape-clusters that are both white and black, the Greeks advise that you should do as follows. If black and white vines are adjacent, split a shoot of each at pruning time, and join them together; by levelling the adjacent eyes of each type, you can unite them as one. Then you will bind them with strips of papyrus. Be sure to smear them with soft moist earth, and water them every three days, until the new leaf-bud breaks open. Then over time, if you wish, you can perform this technique on more shoots.

34. *Hours*

In length of hours, this month accords with November; we calculate them by the following reckoning:

Hour:	1	2	3	4	5	6	7	8	9	10	11
Feet:	27	17	13	10	8	7	8	10	13	17	27

Book 4: March

Field Work

1. *Vines: pruning, grafting and planting*

§1 This month in cold places the pruning of vines will be performed (a subject that we spoke about fully in February [3.12–15]), until such time as there are the first suggestions of budding.

Now is the right time to graft the vines, while they are weeping sap that is thick, not watery. We shall make sure, then, that the trunk being grafted is solid and overflowing with nutritious sap, not in the least dried up by the ravages of age or injury. Next, the scions to be inserted in the decapitated vine[1] should be solid, round and eyed with several close-growing buds. §2 However, three eyes will suffice for the grafting. So the twig should be scraped for a distance of two fingers, with the bark left on one side. Some people do not approve of the pith being laid bare, and scrape only lightly, so that the cut can gradually come to a point, and the part with bark can fit against the bark of its new mother. The lowest eye should be inserted in such a way as to be in snug contact with the trunk. This eye should look outwards, and should be secured by a soaked willow band and on top by mud mixed with chaff; and it should be protected by some kind of covering from being shaken by winds and dried by sun.

§3 When the heat of the season begins, a little water should be splashed frequently from a sponge on the binding around evening, to sustain and refresh it against the violence of the scorching weather. But once the shoot has burst open and put on some growth, it must

---

1. i.e. the vine has been cut down close to the ground, and the scions are to be inserted 'in the trunk': see *Grafting Methods*, 3.17.

be fastened to a support cane, to prevent any movement from shaking the growing twig at this delicate age. When there is a certain amount of more substantial growth, the fastenings should be cut away, to prevent the shoot in its tender adolescence from being strangled by the hard constricting bond.

§4 Some people insert the scions in the vine below ground level, after digging out the soil to a depth of half a foot, and let them benefit from the replacement soil heaped round them; this is a further aid to the new cuttings, in addition to sustenance from the nurse-vine. Some claim that grafting is better done near ground level, because the scions have more difficulty taking at a deeper level.

Vines should be planted up to mid-month or the equinox in cold places, either in dug-over ground or trenches or holes, as described above (3.9).

2. *Meadows and fields*

In cold places meadows should be cleaned now, and reserved for hay. In chilly places it will be appropriate to open and plough over rich hillsides and boggy fields; also to redo the fallow land ploughed up in January.

Sowing of crops

3. In hot dry regions we shall sow *millet* and *Italian millet*. They want a light, loose soil; they come up not only in grit but in sand too, provided that they are sown in damp weather and in well-watered ground, since they abhor dry or clayey land. They must be regularly freed of weeds. Five *sextarii* will fill an area of a juger.

4. We should sow both kinds of *chickpea*[2] now, in very fertile land and in damp weather. It should be soaked the previous day, so it can germinate quicker. A juger will be sown with three *modii*. The Greeks say that chickpea grows big if warm water is poured over it the previous day; that it also loves maritime locations; and that it comes earlier if sown in autumn.

2. Columella 2.10.20 identifies these two cultivated varieties as *arietillum* (so called because of a resemblance to a ram's head, according to Pliny 18.124) and *Punicum* (presumably believed to have come from North Africa).

5. We sow *hemp* this month too, up to the vernal equinox, in the manner discussed in February (3.5).

6. *Chickling vetch* is sown now, in fertile soil that has been first- or second-ploughed. It differs from small chickpea only in colour, being dusky and blacker. We shall fill a juger with four, or three, or even two *modii*.

7. *More on vines*

§1 This month we should start breaking up the soil around a new vine-planting; this should be done regularly from now on, at the start of each month till October, not just because of weeds, but to prevent the hardened ground from constricting the plants while still tender. Grass roots, which are very harmful to the vines, should be pulled up.

§2 Now in cold places the task of digging the vineyards should be undertaken,[3] and the vines should be staked and tied up. (We should fasten a new vine with soft ties, since harder ties cut the tender growth.) A substantial stake should be set for bigger vines, a slender one for smaller vines. Because of the problem of shade it should be set on the north, the cold side, at a distance of four fingers or half a foot from the vine, so the soil can be dug all round the vine.

§3 *To renew an old vineyard.* Some people dock old vines at some height from the ground with the intention of renewing them; but this is unsound, for the quite large wound usually rots as a result of sun and dew. So renewal should be done this way. First it will be ablaqueated fairly deeply, till its root-node is visible. Then it is cut off below ground-level above the node, so that it will be covered and have nothing to fear from cold or sun. This should be done if the vine is of excellent quality and deeply rooted; otherwise it is better to graft it with twigs of choice stock. All the above-mentioned tasks we shall carry out at the first of the month in hot locales, but after mid-month in cold ones.

§4 *To revive sickly vines.* If vines are sickly or their fruit dries up, you will dig around them and pour on aged urine. Or again, apply

3. This is the cultivation between the rows of vines, scheduled by Palladius for February in warm places (3.20) and early April in very cold places (5.2.3).

ash of vine-cuttings or of oak, mixed with vinegar; or cut them close to the ground and revive them with manure, and leave the stronger shoots that come up.

§5 *To cure vines damaged by the hoe.* When a vine is damaged by a hoe or tool, smear the wound, if it is at ground-level, with sheep or goat manure; then be sure to bind it, after the earth has been mixed and dug around the vine. If the damage is on the root, mix in liquid manure when covering it up.

8. *Olive trees*

§1 *If olive trees are struggling*, unsalted *amurca* will be poured now around their roots. For very large trees, as Columella says, six *congii* are sufficient, for middle-sized trees four *congii*, and for others in proportion.[4] Some people lay down bean husks, two baskets for a fairly large tree; others pour what they reckon to be a suitable quantity of aged human urine on the trunk and straightway (especially in dry places) make a 'mortar' for the tree,[5] after covering the trunk.

§2 *If an olive is sterile*, you will bore it with a Gallic augur. Then you take two branches of the same size from the south side of a fruitful tree, and insert them firmly in the hole from each side; cut off what protrudes, and be sure to cover it with mud mixed with chaff. But if they are flourishing yet have no fruit, implant a stake of wild olive, or a stone, or stakes of pine or oak, in their roots.

Other tasks

Now too will be a suitable time to hoe the grain crops again, for those who follow that practice. In cold places, now is the time when nursery beds (discussed in February)[6] for berries and seeds should be established, and cultivation of rose beds[7] should be completed at the start of the month.

---

4. Columella 11.2.29.
5. i.e. shape the earth around the tree into a bowl, to collect rainwater.
6. Nursery beds are defined 3.10.1; used for planting berries, 3.23.2.
7. i.e. the tasks described under February (3.21).

9. Gardens

Cardoon

§1 Now is the best time for gardens to start receiving cultivation. March is the month for sowing cardoon. It likes well-manured loose soil, though it can grow better in rich soil. Against moles it is helpful to plant it in hard ground, so the earth cannot be tunnelled so easily by these harmful creatures. §2 Cardoon should be sown on a waxing moon in a prepared area, the seeds half a foot apart. Care must be taken not to set the seeds upside down, for they will produce weak, crooked, tough stalks. They should not be pushed deep into the soil, but grasped with three fingers and sunk until the soil reaches the first finger-joints. Then they should be lightly covered, continually freed of weeds until the plants have gained strength, and watered if hot weather intervenes. §3 If you break off the tips of the seeds, they will have no prickles. Again, if you moisten their seeds for three days with laurel oil or nard or balm of Gilead, or juice of rose or mastic, then dry and plant them, they will grow with the same flavour as the unguent that the seeds absorbed. Each year, of course, slips should be removed from the stock, so that the mother plants are not exhausted and the offspring can be planted out in different areas. Mind, they should be pulled off with some part of the root. As for those you keep to collect seed, you should free them of all shoots and cover them over with a tile or bark; §4 for the seeds tend to be killed by sun or showers. Against moles it is helpful to keep cats often in the middle of the cardoon beds. Many people keep tame ferrets. Some fill in their entrance-holes with red ochre and juice of wild cucumber. Some open several apertures near the moles' dens, so they are frightened by the entrance of the sunlight and run away. Many set snares in their entrances with loops of animal hair.

Other seedings

§5 This is also the right month for us to sow *garlic* and *ulpicum* and *onions* and *summer savory* in cold places, also *dill*. Now is also an excellent time to sow or plant out *white mustard* and *cabbage*; *mallow* too is sown. *Wild radish* and *oregano* are planted out. *Lettuce* and *beets* and *leeks* and *capers* can be sown, also *Colocasia*, *savory* and *cress*. Some people also sow *endive* and *radish* now, for use in summer.

§6 Now *melons* should be sown, fairly well spaced. The seeds should be two feet apart, in well-worked or trenched ground, preferably sand. The seeds should be soaked for three days in honeywater and milk, and then planted after being dried: this will make them sweet. But if their seeds are tucked for many days among dry rose-leaves, they will be scented.

Cucumbers

§7 Now too cucumbers are sown, well spaced, in furrows one and a half feet deep and three feet wide. Between the furrows you leave an uncultivated space of eight feet, where they can ramble. They are helped by weeds, and so do not need the hoe or weeding. If you soak the seeds in sheep's milk and honeywater, they will grow sweet and white. They become long and tender, if you place water in an open vessel two palms' width below them; they are made so through hurrying to reach it. §8 They will be seedless if their seeds are first greased with savin oil and rubbed with bruised leaves of the herb called 'gnat'.[8] Some people insert the cucumber flower with the end of its vine in a cane, after boring through all its joints; there the cucumber grows, stretched to an uncommon length. It fears olive oil so much that if you plant it nearby, it bends round like a hook. Whenever there is thunder, it turns around as though in fear. §9 If you enclose the flower on its vine in an earthenware mould and fasten it there, then whatever kind of face – human or animal – the mould represents will be shown in the form of the cucumber. All this is stated by Gargilius Martialis. Columella says that if we have brambles or fennel in a sunny well-manured place, we should cut them down near the ground after the autumn equinox, hollow them out with a wooden spike, insert manure within the pith and add cucumber seed; the fruit growing from here cannot fail even in cold weather.[9]

---

8. An unidentified plant, to which Pliny 19.68 also attributes this power of making cucumbers seedless.

9. Columella 11.3.53, who in turn attributes the advice to Bolus of Mendes in Egypt. Gargilius' treatment of cucumbers was presumably part of his lost work on vegetable gardens.

Asparagus

§10 Around the last of this month we shall sow asparagus, in a rich, moist, well-worked place: after small grooves have been made in a straight line, two or three seeds should be placed in each, half a foot apart. Then the ground should be covered with dung, and weeds repeatedly pulled out, or else straw should be thrown on top for the winter, to be removed in early spring. From here the asparagus will grow in three years.[10]

A prompter method, however, is if you plant asparagus 'sponges', which will provide a yield quickly. §11 They are created as follows. After mid-February, in a rich well-manured place, you set in each trench as much asparagus seed as you can hold in three fingers, and cover it lightly. As these grow together an intertwined root will be formed, called a 'sponge'. (But this too involves delays, for after being nurtured for two years in its own nursery bed with dung and constant weeding, it will then be transplanted after the autumn equinox, and will yield asparagus in spring. It will be more convenient to buy these sponges, rather than nurture them over a long waiting period.) At any rate, we shall plant them out in furrows – in the centre of the furrows if the place is dry, but on the tops of the ridges if moist. Moisture should only wet the asparagus sponges in passing, not stand. §12 We should snap off the first asparagus produced, not pull it away, to avoid shifting the sponge while it is still unsteady; but in following years it should be pulled away, to open the germinating eyes, since if you continue to snap it off, the points which are usually productive will be blocked by the asparagus root remaining in place. They will produce in spring, and in autumn you will leave the one from which you intend to take seed. Afterwards you burn its stalks, then close to winter you will spread manure and ash on the sponges.

Rue

§13 This month rue is sown, in sunny places – content with just a sprinkling of ash. It requires fairly elevated spots from which moisture drains away. If you plant its seeds still enclosed in their pods, you will have to set these individually by hand; but if the small seeds are already separated you will broadcast them thinly and rake them

10. Counting inclusively, i.e. with the year of sowing as the first year.

under to cover them. The stems that grow from enclosed seeds will be stronger, but slow to grow. The twigs, pulled off with part of the bark in springtime, will take as an alternative to seedlings, but if the whole plant is moved it will die. §14 Some people bore a hole in a bean or bulb, insert the rue twigs and bury them like that, to be nurtured by borrowed strength. People even execrate them with curses, and in particular they plant them in soil of loose brick-clay, which is absolutely certain to help them. But they claim that stolen rue will grow better. It is happier in the shade of a fig tree. It wants weeds be pulled up, not dug out. It fears the touch of an impure woman.[11]

More seedings

§15 *Coriander* is sown from this month right through October. It loves rich ground, but does grow in lean soil too. Older seed is considered better. It is keen on moisture. It will grow well if sown alongside any vegetable at all.

§16 This month *gourd* should be sown. It loves soil that is rich, moist, well-manured, loose. A notable point about gourds is that seeds which grew in the neck produce long thin ones; those which were in the belly make fatter gourds; those which were in the base make broad ones, if planted head-down. When they start to mature, they should be supported with props. Those kept for seed should be left hanging on the vines till winter, then removed and placed in the sun or smoke. Otherwise the seeds rot and die.

§17 This month *blite spinach* is sown, in any kind of cultivated soil. This vegetable does not need either weeding or hoeing. Once grown it will replace itself for many generations by dropping seed, with the result that it can hardly be suppressed even if you should wish to.

Now too *creeping thyme* is started by seedlings or seed, better with age. It will leaf up more profusely if grown near a pond or tank or the edging of a well.

Anise too and *cumin* sow well now. They grow better in more fertile places, but in others too if helped with moisture and dung.

---

11. i.e. during menstruation.

10. Fruit Trees

Pomegranate

§1 In temperate places we shall start pomegranates in March or April, but in hot dry places in November. This tree loves chalky lean soil, but it grows in rich soil too. A hot region is suited to it. It is started with slips pulled from the mothers' root. But although there are many methods of starting it, the best is this: a branch is cut, a cubit's length and the thickness of a hoe handle; it is trimmed at each end with a sharp pruning-blade, and planted in a hole at an angle, like a vine-cutting. (First, however, it should be smeared with pig manure on the top and bottom.) Or in uncultivated soil it should be driven with a mallet, so as to sit at a lower level. §2 It will grow better if the branch to be planted is taken when the mother tree is already in bud. But when planting in a hole, if the planter sets three stones in the root,[12] he will ensure that the fruit does not split. Care must be taken not to plant the growth upside-down.

It is believed that they become tart if watered repeatedly; in this species drought results in both sweetness and a plentiful crop. But some moisture should be provided, to prevent a surfeit. §3 It should be dug around in autumn and spring. If the fruit is tart, a modicum of asafoetida ground up with wine should be poured on the tree tops, or else the roots should be ablaqueated and a fir peg set into them. Others dig in seaweed around the roots, and some people mix ass or pig manure with it. If it does not hold its flower, you will dilute aged urine with an equal quantity of water and pour it on the roots thrice a year. An infusion of an *amphora* will suffice for one tree. Or you pour on unsalted *amurca*, or place seaweed on the roots and water twice a month, or else you must encircle the trunk of the tree when in flower with a lead hoop, or wrap a snakeskin round it.

§4 If the fruit splits, you place a stone in the middle of the tree's root, or sow squill around the tree. And if, while the fruits are hanging, you twist them on their stalks as they are held on the tree, they keep up to a full year without spoiling. If they are troubled by worms, you touch the roots with ox gall and they die straightway;

12. This presumably refers to a slip not a 'branch' (i.e. a truncheon), since the latter would not have a root.

or if you clear off the worms with a bronze nail, their reproduction will be curbed; or else ass urine mixed with pig manure will repel the worms. A frequent infusion of ash with lye around the trunk of the pomegranate will make the trees vigorous and fruitful.

§5 Martialis states that white seeds are produced if you mix argil or clay with a fourth part of gypsum and add this soil preparation to the roots for a full three years. He also says they grow to a wonderful size if a clay pot is buried near the pomegranate tree and a flowering branch is enclosed in it, fastened to a stake so as not to spring back. Then the pot should be covered and protected against the entry of water. When opened in autumn, it will reveal fruit of the same size as itself. §6 The same author states that many fruits grow on a pomegranate tree if an equal mixture of juice of spurge and purslane is painted on the tree's trunk before it buds. Grafting can be done, so it is affirmed, by fastening branches together, with the pithy centres of the two split branches making a union.[13] It can be grafted only on itself, near the start of April, at the last of March. When the trunk is cut a very fresh scion needs to be inserted straightway, lest any delay should dry up the small amount of moisture in it.

Pomegranate: preservation, wine-making

§7 Pomegranates keep if you apply pitch to the stalks and hang them up in rows. Alternatively they should be picked unblemished and submerged in boiling seawater or brine, so as to absorb it. They should be dried for three days in the sun, but without being left outdoors at night; then they should be hung up in a chilly place. When you want to use them, you will soak them in fresh water the day before. They are said to rival the fresh fruit. §8 Likewise if they are buried in chaff, spaced so as not to touch each other. Likewise, a long trench is dug and a piece of bark is obtained of the same width, to which the fruits are fastened by their own sharp twigs. Then the bark is inverted and placed over the trench, so it protects the fruit hanging below it from moisture, while not allowing any contact with the earth. Likewise if they are coated with fine clay and, after it has dried, hung in a chilly place. §9 Likewise if a jar is dug into

13. The procedure seems to be similar to that described at 3.33, for bonding shoots from two adjacent vines.

the ground in the open and filled halfway with sand, and the fruits picked with their stalks are each pushed into a single cane or elder stem, and set in the sand like this, with the fruits not touching and four fingers clear of the sand.[14] This can also be done under shelter, in a three-foot hole. It is conducive to keeping if the fruits are picked with a fairly long branch. §10 Alternatively the fruits are suspended in a jar half full of water, without touching the water, and the jar is closed so breezes cannot force their way in. Likewise, they are buried in barley in a storage vessel, arranged so as not to touch each other; the vessel is covered on top.

You will make pomegranate wine as follows. You place the ripe pomegranates, carefully cleaned, in a palm basket, crush them in a screw press, and boil the juice down gently to half the quantity. When it cools, you store it in vessels sealed with pitch and gypsum. Some people do not heat the juice, but mix it with honey, one *libra* per *sextarius*, and keep it in the aforementioned vessels.

Citron: propagation

§11 The citron tree is started in March, by many methods: seeds, branch, cutting, truncheon. It loves ground with a rather open texture, a hot climate and continual moisture. If you want to start it from seed, you will do so as follows. You will dig the earth to two feet, mix in ash, and make small beds, such that water can run off through channels on each side. In these beds you will open up a palm-deep hole by hand, set there three seeds side by side with their points turned down, cover and water daily. They will come up quicker if you use warm water to help them. §12 Once they have sprouted, nearby weeds must always be cleared away. A seedling can be transplanted from here at three years.

If you want to layer a branch, you should not bury it deeper than a foot and a half, lest it rot. Starting a tree with a truncheon is more advantageous. It is made the thickness of an hoe handle, a cubit in length, trimmed at each end, with knots and thorns cut away, but with the topmost buds intact, so the promise of future shoots can swell in them. §13 Particularly diligent people also smear each end

14. i.e. the stalks of the pomegranates are inserted in the tops of the canes or stems, and these are then pushed into the sand.

with cattle manure, or wrap it in seaweed, or cover each end with well-worked fine clay and plant it like that in trenched ground. A cutting can be both thinner and shorter. It will be planted like a truncheon, but it should protrude two palms' height from the ground, whereas a truncheon is buried completely. In spacing it does not require particularly large gaps. It should not be planted in association with other trees.

§14 It enjoys hot but well-watered places, especially coastal places where moisture abounds. But if anyone compels this species to grow in a cold region, he should plant the tree in a place that is protected by walls or else faces south. Even so, in the winter months he should cover and protect it with straw from the fields; when summer shines again, the tree can be restored to the air, naked and safe. §15 In very hot regions it is planted in autumn too, by cutting or truncheon; in very cold regions I myself have planted it in July and August, quickened it with daily watering and induced it to bear fruit and grow large.

Citron: cultivation and preservation

People believe that citron is helped if gourds are sown nearby; their vines, too, when burnt provide ash that is useful to citron trees. §16 They enjoy repeated digging, which makes the fruit larger. We should prune them very rarely, except for dried-up growth. Citron is grafted in April in hot places and in May in cold ones, not under the bark but in the trunk which has been split right by the roots.[15] It is grafted both on pear and, as some say, on black mulberry; in any case, the scions once inserted must by all means be protected above by a wicker basket or earthenware vessel.

Martialis states that in Assyria this tree never stops fruiting. I learnt this by experience in the Neapolitan territory in Sardinia[16] on my farms, which have a warm soil and climate and abundant moisture: the fruits succeed each other in a kind of series, with unripe ones taking over from ripe ones and those in flower following those at the unripe stage, as nature furnishes itself with a kind of cycle of continuous fertility. §17 It is said that the pith changes from bitter

15. Again see *Grafting Methods*, 3.17.
16. This territory lay on the south-west coast of Sardinia.

to sweet if seeds for planting are soaked for three days in honeywater or (better still) in sheep's milk. Some people bore a hole in the trunk, slanting up from the base and not reaching the other side, in February. From this they allow moisture to flow until the fruit is formed; then they fill the hole with mud. They say the result is that the middle of the fruit becomes sweet.

§18 Citron can actually keep almost a whole year on the tree; better, if enclosed in small pots. If you want to gather and keep them, you will need to pick them at night, with the moon hidden, on their leafed branches, and lay them out separately from each other. Some people place each in an individual pot, or coat them with gypsum and store them in rows in a shaded place. Most people keep them covered in cedar sawdust or chopped straw or chaff.

Medlar

§19 Medlars are happiest in places that are hot but well-watered; however, they do grow in cold places too. They prefer fertile grit or gravelly soil that has sand mixed with it, or argil with rocks. It should be started by cuttings in March or November, but only in soil that is well manured and well worked; each end of the cutting should be covered with dung. Its growth is extremely slow. It likes to be cut back and dug around, and to be refreshed frequently with small amounts of water during dry periods. It is started by seed too, but this means being hopeful for a longer period.

§20 If it is attacked by worms, they should be cleared out with a bronze spike and sprinkled with *amurca* or aged human urine or quicklime (but sparingly because of damage to the tree) or water from boiled lupine (but this is thought to make the tree sterile). If dung and the ash of vines are poured on the roots together, they render the tree fertile. §21 If ants are troublesome, they will be killed by red ochre blended with vinegar and ash. If the fruit falls, a snippet cut from its root should be fastened in the centre of the trunk.

It is grafted in February on itself or on pear or apple. The scion, mind, must be taken from the middle of the tree, for one from the crown is unsound. It must be grafted onto the split trunk, for the hungry leanness of the bark will give no sustenance.

§22 Medlars are gathered for keeping while not yet soft. They will last a long time on the tree, or in pitched jars or hung up in rows or (according to some) stored in vinegar-water. They should be gathered on a fine day at midday and buried in chaff, spaced out to prevent spoilage through contact with each other. Or they may be gathered half-ripe with their stalks, soaked in salt water for five days, and then dropped in *sapa* so as to swim in it. They are also kept in honey, but only if you collect them when very ripe.

Figs: propagation and cultivation

§23 Rooted fig seedlings should be planted in November in hot places, in February in temperate ones, but better in March or April in cold ones – at the end of April if you are planting a cutting or branch-tip,[17] when livelier sap has flowed into it. When the seedling is set in the hole, stones should be placed beneath it; the soil near the root should be mixed with dung. If the place is cold, the tops of seedlings should be protected from the cold with split internodes of cane. §24 If you want to plant a branch-tip, you will cut a three-forked branch two or three years old from the south side, and bury it in such a way that the tips, separated by the soil lying between them, look like three individual slips. A cutting will be planted like other species; after lightly splitting it at the bottom we shall insert a stone in the cleft. I myself have planted out large fig saplings in trenched ground in Italy in late February or March; the same year, above and beyond their success in taking root, they produced fruit, as if making a recompense. §25 We should choose seedlings that have protuberant nodes close together; those that are shiny and have eyes separated by long gaps are believed to be sterile. If you first nurture a fig seedling in a nursery bed and transplant it into a hole in due time, it will produce superior fruit. Some people declare that it is very helpful if we split a squill bulb, place the fig seedling between the halves, tie them together and plant. It likes deep holes, wide spacing, a type of earth that is hard and light – and dry, for the sake of acceptable flavour in the fruit.

§26 It grows even in harsh rocky places, but it can be started in almost any kind of location. Those growing in montane or cold places

17. A cutting taken from the top of the tree.

cannot last to the dried-fig stage, since they have less juice; they are used green, having larger size and sharp flavour. Those growing in flatlands and hot places are more luxuriant, and also keep well when dried.

§27 If we wished to count the varieties, it would be an immense task; suffice it to say that cultivation is the same for all. There is this difference, that among Carian figs the white ones keep better. In exceptionally cold places we should plant early-ripening figs that come quickly, so this type can arrive before the rains; but in hot and sultry places, the late-maturing kind. It enjoys repeated digging. In autumn it will be helpful if you bring in some manure, particularly from the aviaries. You should cut back any decayed or ill-formed growth you find, and prune in such a way that the tree is directed outwards and expands to the sides.

§28 In moist places figs are dull-flavoured. Against this, after cutting around the roots, a certain amount of ash should be sprinkled on. Some people grow a wild fig in the fig orchards, to avoid the need to hang up the same fruit in each tree as a remedy. In June around the solstice fig trees should be 'caprificated', i.e. immature figs from a wild fig tree should be hung up, threaded on a string like a garland. If these are not available, a twig of wormwood is hung up, or the callus found on elm leaves, or else a ram's horns are buried near the tree's roots; or the tree trunk is scarified at a swelling point, so the moisture can flow out.

Figs: remedies and grafting

§29 To prevent it suffering worms, we shall plant a branch of terebinth or cutting of mastic upside-down with the fig seedlings. Bronze hooks should be used to remove worms from a fig tree. Some people mix *amurca*, others aged urine, into the ablaqueated roots. Others smear bitumen and olive oil, or quicklime alone, on the worms' hiding places. If ants are a nuisance, red ochre mixed with butter and liquid pitch should be applied around the trunk. Others declare that against worms a raven-fish [18] should be hung up in the tree. §30 If the tree drops its fruits as though sickly, some people smear it with red ochre or unsalted *amurca* mixed with water, or hang up a river crab together with a branch of rue, or seaweed or a bundle

18. A species of fish found in the Nile, so called for its black colour.

of lupines, or bore the root and fasten a wedge in it, or slit the tree's skin frequently with an axe. When fig trees begin to produce leaves, to make them bear a quantity of plump fruit we cut off the topmost growth as the buds start to open, or else just the top-growth coming from the middle of the tree. §31 If you want to make an early tree late-bearing, knock off the starting figs when they are the size of a bean. To make a fig tree ripen early, smear the fruit with the juice of a longish onion, mixed with oil and pepper, when the young figs start to have a reddish tinge.

We should graft the fig in April, in the bark or, if the trees are young, in the split trunk-wood, which should straightway be covered and bound up, so the wind cannot get in. §32 They take better if the plants are cut off close to the ground before grafting. Some people even graft in June. We should choose a year-old scion, for an older or younger one is believed to be useless. Bud-grafting of a fig can be done in dry places in April, in moist ones in mid-July, in warm ones in October. Grafting is done onto wild fig, black mulberry or plane, both by 'eyes' and by scions.[19]

Figs: preservation, etc.
§33 Green figs can be kept either in honey, arranged so as not to touch each other, or enclosed individually inside a green gourd, in spaces hollowed out for each one and closed again with the square that was cut out, the gourd being hung up where there is no fire or smoke. Some people gather fresh unripe figs with their stalks, enclose them in a new earthenware vessel not touching each other, and let the vessel float in a jar full of wine. §34 Martialis says Carian figs are kept by many methods, though one system is sufficient. So we should keep them by the following method, which is used throughout Campania. The figs are spread out on wickerwork mats until midday, and poured back into a basket while still soft. Then the basket is placed in an oven heated to bread-baking temperature, with three stones placed beneath it to prevent it catching fire, and the oven is closed. Once the figs are softened by cooking they are placed, still hot, in a well-pitched earthenware vessel between layers

19. The method of grafting called bud-grafting, using buds or 'eyes', is described at 7.5.

of their own leaves, tightly packed, and the vessel is carefully sealed with a lid. §35 If you cannot spread out the wicker mats because of copious rains, you place them under shelter, raised half a foot from the ground, and have ashes set beneath them to heat them in lieu of the sun; the figs, split as they are, should be turned at intervals, so their skins and pulp can dry. Then they should be folded over and kept in crates or boxes. Other people split the figs when moderately ripe and spread them on wicker mats to dry all day in the sun; they bring them under cover at night.

§36 Now is a useful time for branch-tips of fig trees to be planted, when the buds are swelling, to make young plants if you do not have an abundant supply of them. To make a single tree display variegated fruit, you fasten together two branches, from a black and a white tree, and bend them in such a way that they are forced to intermingle their buds. They are planted like this, and helped with dung and moisture. When they begin to grow, stick the germinating buds together with some kind of fastening. Then this melded shoot will produce two colours, diversified in unity and joined in diversity.

Fruits discussed in other months

§37 Now both pear and apple are grafted and started; quince and plum are grafted; service-berry is planted, also black mulberry on 24th March;[20] pistachio is grafted, and pine seed is scattered in cold places.

Livestock

11. *Securing oxen, bulls, cows*

§1 This is the month for securing oxen. Whether they are taken from our own herds or purchased, now is a suitable time for obtaining them, for the following reasons: since they are not yet stuffed with seasonal plenty, they cannot conceal their own faults or the seller's tricks; also they do not have the confidence of full-fed strength, which would make them headstrong in resisting taming.

20. For the date cf. 3.25.28.

§2 These are the points to be watched for in oxen, whether they are to be obtained from our own herd or another. The oxen should be young and blocky, with massive limbs, a solid body, protuberant muscles everywhere, large ears, a broad curly-haired forehead, blackish lips and eyes, sturdy horns of crescent form with no defective crookedness, nostrils that are wide open and splayed back, a well-built muscular neck, large dewlaps flowing down to the knees, a big chest, huge shoulders, not a small belly, lengthy flanks, broad loins, a straight flat back, legs that are solid, sinewy and short, big hooves, a long coarse-haired tail, dense short hairs all over the body, and preferably red or dusky in colour.

§3 We shall do better to obtain oxen from nearby areas, since there will be no change of soil or atmosphere to trouble them. Or, failing this, we should transfer them from like places to like. Before all else we must ensure that the animals obtained are equal in strength for pulling, lest the vigour of the stronger one should be the other's undoing. In temperament the following should be considered: they should be keen, gentle, submissive to prompting by shout or blow, eager for feed. If the nature of the region allows, there is no better feed than green fodder; but where that is lacking, one will feed them as the supply of fodder and the onset of tasks dictates.

§4 Now too a person whose heart is set on building up a herd will obtain bulls, or raise them from a tender age, on the basis of these points: they should be tall, big-limbed, of middling age (or on the younger side rather than declining into old age), with a fierce face, small horns, a massive muscular neck, and compact belly.

§5 Now is also the prime time to secure cows. We shall choose those of very tall form, long body, large roomy uterus, broad forehead, big black eyes, handsome horns (preferably black), a hairy ear, very large dewlaps and tails, compact hooves with small black legs, and preferably three years old, since the offspring they produce up to their tenth year are more satisfactory, but before their third year they should not be bred. §6 A diligent farmer will make it his regular duty to remove older cows, gather young ones and assign barren ones to the plough and other tasks.

The Greeks state that if you want to produce males, the bull's left testicle should be tied up in breeding, and if females, the right.

They add that for a lengthy prior period the bulls should be denied access, so as to bear down more eagerly, when the time comes, on the causes of their deferred ardour.

§7 For cattle we should provide sunny maritime locations in winter and shady cool ones in summer, especially in the mountains, since they are better satisfied by shrubs and by the herbage growing among them. Though they are rightly pastured near rivers because of the pleasantness of those places, their reproduction is aided by warmer water; consequently it is more suitable to keep them where rainwater forms warmish pools. §8 Nevertheless, this species of livestock does tolerate cold, and can easily winter in the open. It is appropriate to create rather spacious folds for them, to avoid harm to the pregnant cows. Their sheds are usefully floored with rock or gravel or sand, sloping somewhat downwards so the moisture can drain out, and facing south on account of icy winds, against which there should be some protective barrier.

12. *Taming oxen*

§1 At the end of this month three-year-old oxen should be tamed: they cannot be successfully tamed beyond five, since they are hardened by age and resist. So they should be caught and tamed straightway; in fact earlier, while still young and tender, they should be gentled by frequent handling. New oxen will need to have a spacious shed, such that the area in front of it is not constricted, and when led out they are not unsettled by hitting some obstacle. §2 In the shed itself beams should be fastened horizontally seven feet from the ground, so the untamed oxen can be tied to them. Then you pick a day, free of bad weather and any hindrances, on which they can be caught and led to the shed. If their wildness is too great, they should be softened up by being kept tied and hungry for one day and night. Then, while speaking to them in a coaxing way and enticing them by offering food, the oxherd (approaching not from the side or back but from the front) should stroke them and handle the nose and back, sprinkling them occasionally with unmixed wine. But he should take precautions that the ox not contact anyone with his hoof or horn, since if he perceives at the outset that it has worked to his advantage, this fault will take hold. §3 Then, when they have been softened up, rub their mouth and palate with salt, and put down their gullets one-pound cakes made with extra-salty lard, and a *sextarius* each of wine, using a horn to pour it down their throat. This treatment will completely relax their angry wild nature within three days. Some people yoke them together and teach them to tackle light loads. It is useful, if they are being prepared for ploughing, to train them on soil that has previously been tilled, so the new task does not bruise their necks while still tender.

§4 A more expeditious method of taming, however, is to yoke a wild ox to one who is gentle and strong; he will easily be constrained into performing any duty, with the other demonstrating how. If after taming he lies down in the furrow, he should not be dealt with by fire or blows; rather, when he lies down, his feet should be tied so that he cannot walk or stand or graze. When this is done he will be worn down by thirst and hunger, and lose this fault.

13. *Horses, mares, and foals*

§1 This month, after being conditioned and well fed, the stallions should be let in to the mares of good stock, and gathered back to their stalls again after the females have been impregnated. But we should not introduce an equal number to all; rather we should assess the strength of each stallion, and then set few or numerous mates to him. This policy will ensure that the stallions last for more than a short career. However, even if a male is in his prime and of settled strength and conformation, we should not set more than 12 or 15 to him; to others in accordance with their strength.

§2 We shall look for four things in a stallion: conformation, colour, merit, beauty. In conformation we shall aim for the following: a massive, solid body, height matched with sturdiness, a very long flank, large round buttocks, a broad chest, the whole body knotted with thick muscle, a foot that is dry and solid with a fairly high hoof of hollow horn. In beauty the elements are as follows: the head should be small and dry, with the skin almost attached directly to the bone; short, sharply defined ears; big eyes; flared nostrils; quite profuse mane and tail; a solid, firm roundness in the hooves. §3 The colours are these in particular: bay, golden, *abineus*,[21] red-brown, myrtle, deer-brown, chequered dun, dappled white, pure white, dull black. Of second quality is a beautifully pied colouring with a mixture of black or *abineus* or bay; grey combined with some colour; a variegated foam colour; a rather dull mouse colour. But in stallions we should choose particularly those of a single distinct colour. Others should be rejected unless the greatness of their merits excuses their faulty colouring.

§4 The same points should be considered in mares, especially that they should be long and large in the belly and body. But this is a consideration in animals of good stock. As for the other mares, the stallions can be let out with them all year without restriction throughout the pastures, and impregnate them there. Mares naturally bring their pregnancy to term somewhere in the twelfth month. Something to be observed for stallions is that they should be separated by some intervening space, to prevent harm from their rage at each other. For these animals we should choose very rich pastures that are sunny in winter, cold and shaded in summer, and not on such soft ground that the firmness of their hooves never encounters any rough surface.

§5 If a mare will not accept a male, suffusing her genitalia with ground-up squill will kindle her desire. Then the mares once pregnant should not be stressed, nor endure hunger and cold, nor be crowded together in a constricted place. Mares of good stock, and those raising males, should be bred only every second year, so they can pour the strength of their copious pure milk into their foals; the others can be bred without restriction. §6 The age for a stallion to start should be the beginning of his fifth year; but a female will properly conceive even as young as two, since after her tenth year any offspring born from her will be feeble and sluggish.

The foals when born should not be touched by hand, as repeated touching harms them. As much as is practical, they should be protected from cold. In foals, with allowance for age, those points should be considered which show signs of good quality, and which are to be looked for in the sires or dams, as I have instructed. Liveliness,

21. *Abineus* occurs twice in this section but nowhere else in extant Latin, and its meaning is unknown.

alertness and agility will also serve as pointers. §7 Foals that have passed the age of two should be tamed at this time. Points to be considered are big, long, muscular, well-defined bodies, testicles that are matching and small, and the other things mentioned in regard to the sires; temperament such that they are easily aroused from deep rest or checked without difficulty after a hard gallop.

§8 Their age is appraised as follows. At two years six months the upper middle teeth fall out. At four years the canines are replaced. During the sixth year the upper molars fall out. In the sixth year those that were first replaced grow level. In the seventh year all his teeth are filled out.[22] From here on, marks of age are not evident. But in animals of more advanced age the temples begin to sink, the eyebrows to whiten, the teeth generally to protrude.

This is the month in which we should castrate all quadrupeds, especially horses.

14. *Mules and donkeys*

§1 If anyone desires to propagate mules, he should select a mare of large body, solid bones and excellent conformation, looking not for speed in her, but strength. An age from four to ten will be suitable for this kind of breeding. If the male donkey shows no interest in the mare when put in, our recourse is to show him a female donkey first, till the relish for mating is stimulated. When she is removed, his aroused desire will not spurn the mare; carried away by this enticement, he will consent to the mixing of his race with another. §2 If in his frenzy he hurts the mares offered to him by biting, he should be gentled somewhat by work. Mules can be bred from a stallion and female donkey, or from a wild ass and mare. But no animal of this kind is superior to one born of a donkey sire. Mating a wild ass with a female donkey will produce useful stallions, such as will re-introduce agility and courage later, in the generation to follow.[23] §3 A donkey stallion should be as follows: of ample, solid, muscular body, well-knit strong limbs, and black or preferably mouse-grey in

22. This material comes from Columella 6.29.5. The reference of 'grow level' and 'are filled out', at least in Columella's source, was probably to the smoothing-out of the depressions or 'cups' in the upper surface of the incisors.
23. i.e. in the grandsons of the wild ass.

colour, or red. But if he has hairs of a different colour on his eyelashes or ears, he will generally give rise to mixed colour in his offspring. He should not be let in when younger than three or older than ten. §4 A female mule should be weaned from her dam at one year old and pastured on rough mountainsides, so she will be toughened at a tender age and think nothing of the toil of travelling. But the smaller donkey is essential to the farm, since he not only tolerates hard work but also puts up with virtual neglect.[24]

15. *Bees*

§1 This month particularly, bees tend to be afflicted by disease. For after their winter hunger they seek out eagerly the bitter flowers of spurge and elm, which open first; so they incur a loosening of the stomach and perish, unless you are at hand with a speedy remedy. You will provide them, then, with ground pomegranate seeds in Aminean wine, or raisin seeds with Syrian sumach and dry wine, or all these ingredients together, pulverized and cooked in sharp wine. They should then be cooled and placed in wooden troughs. Again, rosemary boiled in honeywater is chilled and this liquid is placed in a cupped tile.

§2 But if they look rough and pinched, sluggish and silent, frequently bringing out dead bodies, you should pour honey heated with powder of gall-nut or dried rose in troughs made of cane. Above all it will be effective to cut away regularly mouldy parts of the combs, or empty cells that the swarm will not be able to fill when reduced to small numbers by some mischance. You should do this with very sharp tools and delicately, to avoid shifting another section of the combs and forcing the bees to abandon their dislodged abode.

§3 Bees are often harmed by their own good fortune. For if the year produces a great abundance of flowers, the bees devote their attention solely to bringing in honey, and give no thought to offspring. With that process of renewal ignored, the current population perishes worn out with toil, and the whole line is destroyed. So when you see

24. Palladius is talking here of a smaller type of donkey (*asellus*) used only for farm work, as distinct from the larger type (*asinus*) used for breeding mules. He is following Columella 6.37.10, 7.1.

an excessive flow of honey as a result of a large and sustained harvest of flowers, you will shut the hive entrance at three-day intervals and stop them going out. In this way they will apply themselves to producing offspring.

§4 Now around the last of the month we must attend to the hives, so as to remove all the rubbish and dirt that has accumulated in wintertime, and the caterpillars and grubs and spiders that taint the usefulness of the combs, and moths that cause caterpillars to be born from their droppings. Then the smoke of burning dry ox dung should be employed, as it is beneficial to the bees' health. This cleaning should be practised frequently right into the autumn. You will carry out all these and other tasks while chaste and sober, and unaffected by the baths or acrid foul-smelling foods or any salted food.

16. *Hours*

For determining the hours, this month agrees with October:

Hour:	1	2	3	4	5	6	7	8	9	10	11
Feet:	25	15	11	8	6	5	6	8	11	15	25

Book 5: April

Field Work

1. *Alfalfa: sowing and care*

§1 This is the month to sow alfalfa in the beds you prepared earlier, as we discussed (3.6). Once sown, it lasts ten years, and can be cut four to six times per year. It fertilizes the ground, conditions thin animals, and cures sick ones. A juger of it is amply sufficient for three horses for the whole year. A *cyathus* of seed covers an area five feet wide by ten feet long. §2 Mind, the seed once broadcast should promptly be buried with wooden rakes, as it is quickly scorched by the sun. After sowing, you must not touch the place with anything iron: wooden rakes should be used to clear the weeds – frequently, so they do not choke the tender alfalfa. §3 The first harvest should be made fairly late, to allow it to drop some seed. Subsequent harvests, however, may be carried out as soon as you wish, and supplied to the livestock. But this new fodder should be supplied rather sparingly at first, for it bloats and creates much blood. After cutting, water frequently. After a few days, when it starts to put forth shoots, weed out all other plants. This way you can harvest six times a year, and it will still be able to last for ten straight years.

2. *Grafting olives*

§1 In temperate places the olive should be grafted now. It is grafted in the bark, like fruit trees, as described above (3.17). In grafting on wild olive, we need to counteract the danger that an olive orchard, grafted and then accidentally burnt, will grow back as the infertile wild olive.[1] So we take the following precautions. First we shall set wild olive branches in the hole where we plan to do the grafting, then refill the holes but only halfway. §2 When the wild olive has taken, we shall graft it at the base (or plant it out if already grafted), and feed the graft by piling the earth just above it. Then, as it grows, we heap on more soil periodically. With the graft hidden deeply in this way, anyone who burns or cuts the tree will not destroy the place from which the cultivated olive can sprout. From the cultivated olive it will retain a fertile capacity to regrow above ground; underground, from its bond with the wild olive, it will retain a fruitful ability to thrive.

§3 Some people graft olive trees on their roots, and when the grafts have taken, they pull them off with part of the root, and plant them out like rooted cuttings. The Greeks advise that olives should be grafted from 25th March to 5th July – lateish in cold locales, and earlyish in hot ones.

Other field tasks

In very cold places the digging of the vines should be completed now before mid-month, plus any tasks left over from March. §4 We also graft vines. The nursery beds made earlier (3.10) should be freed of weeds and dug lightly. In moderately dry places we shall sow millet and Italian millet now. Rich ground, and fields that hold moisture for a long time, should be first-ploughed after mid-month, when their weeds have all grown but have not yet set seed.

3. Gardens

§1 This month too, at month-end with spring nearly over, we can sow *cabbage*. It will serve for stem, since it has missed the time for spring greens.[2]

---

1. i.e. if a cultivated olive has been grafted onto wild stock, and a fire then reaches below the point of the graft, it will destroy the cultivated growth, leaving only the wild stock to regrow.
2. It appears that not only the leaves of cabbage were eaten but also the stem. Kohlrabi is a modern cultivar of cabbage developed for stem.

This is a good time to sow *parsley*, in either hot or cold locales and in any kind of soil, as long as there is constant moisture – though it does not refuse to grow, if necessary, in dry conditions, and it may be sown in almost any month from early spring to late autumn. §2 The same family contains horse parsley (tougher, though, and more bitter), and marsh parsley with soft leaves and tender stalks, which grows in ponds, and rock parsley especially in rough places. Diligent gardeners can have all these varieties. You will make the parsley plants bigger if you enclose as much seed as you can grasp with three fingers in a thin linen strip, and bury it in a shallow trench. In this way the shoots of all the seeds will mesh together in a single dense head. They grow curly if the seeds are first bruised, or if weights are rolled over the growing beds, or the seedlings are trodden underfoot. Older parsley seeds germinate quicker; those that are fresh, more slowly.

§3 This month we shall sow *orach*, if we can water it; also in July or subsequent months until autumn. It loves being drenched with constant moisture. The seed must be buried immediately after sowing; weeds should regularly be pulled out around it. Transplanting is not necessary if it is sown properly; nevertheless it can grow better if planted out with more space, and helped with the 'juice' of fertilizer and water. Mind, it should be cut back with the knife all the time, because in this way it sprouts continually.

§4 *Basil* is sown now. They say it germinates quickly if you drench it with hot water immediately after sowing. Martialis asserts a wonderful thing about basil, that it produces now purple flowers, now white, now red, and that if sown repeatedly from the same seed, it changes now into creeping thyme, now into mint.

§5 This month too *melons* and *cucumbers* are sown, also *leeks*, and at the start of the month we shall set out *capers* and *creeping thyme* and *Colocasia* seedlings. We shall also sow *lettuces* and *beets* and *onions* and *coriander*, and a second sowing of *endive* for use in summer, and *gourds* and *mint* from root or seedling.

4. Fruit Trees

Jujube

§1 We shall plant jujube in April in hot places, but in May or June in cold ones. It likes hot sunny places. It is started from kernels or stock[3] or seedling. It grows very slowly. But if planting seedlings, do so rather in March in soft ground; if starting it from kernels, do so in a hole a palm deep, with three seeds placed upside-down in each hole. Manure and ash should be poured in the bottom and top of the hole, and the emerging seedling should be freed by hand from weeds growing near it. §2 When it is as thick as a finger, it should be moved to a trenched area or into a hole. It likes ground that is not too rich, but close to thin and hungry. During winter it benefits from having a pile of stones heaped around the trunk; these should be removed in summer. §3 If this tree is unhappy, it will cheer up if the trunk is scraped clean with an iron strigil, or if you pour a little cattle dung regularly around its roots. The jujube fruits, gathered when ripe, are kept in a long earthenware vessel which is sealed and placed in a dry spot. Or if you drench the freshly gathered fruits with a sprinkling of old wine, the result is that they are not disfigured by shrinkage and wrinkles. They are also preserved by being cut down on their branches, or wrapped in their own leaves and hung up.

Fruits discussed in other months

§4 Also this month, in temperate places, *pomegranates* are planted in the manner described (4.10.1–2), and are grafted. Now the *peach* can be bud-grafted around the last of the month in the way in which a fig tree is shield-budded, as we said in talking about grafting (4.10.32). This month in hot places the *citron* tree is grafted, as I described above (4.10.16). Now we shall set out slips of *fig* in cold places, keeping to the system described above (4.10.23–25). §5 Now too we should graft the fig in the trunk-wood or under the bark, as I instructed earlier (4.10.31–32), and bud-graft it in dry places. Now the young *palm* plant, which we call the 'head',[4] should be set out in sunny hot

3. Perhaps branches planted as truncheons, or possibly cuttings.
4. Presumably the terminal bud, also called the 'brain' (e.g. Pliny 13.36 and 39); the modern metaphor is 'heart'.

places. This month we can graft *service-berry* onto itself, onto quince or onto whitethorn.

Miscellaneous

5. *Violet-infused oil and wine*

Infuse as many ounces of violet as you have *librae* of oil; you should keep it in the open air for 40 days.

You should infuse five *librae* of violets, wiped clean of all moisture, in ten *sextarii* of old wine; after 30 days, sweeten with ten *librae* of honey.

6. *Livestock*

This is the month when calves are born. Their mothers should be supported with an abundance of feed, so they can contribute their dues in both labour and milk. As for the calves themselves, millet that has been roasted and ground should be mixed with milk, to be provided like mash.[5] This is the time for sheep to be shorn in hot regions, and late-born lambs should be marked this month. Now is the earliest time to put in the rams – and an excellent time too, to ensure that winter will find the lambs already well grown.[6]

7. *Bees*

Finding a swarm

§1 This month we shall search for bees in suitable places. Honey-producing places are indicated by the bees foraging in very large numbers around springs; where they are rarely seen, they cannot produce honey effectively. If they are drinking in large numbers, we shall discover the location of their swarms in the following way.

§2 The first thing is to discover how far away or how close they are. We should take along some liquid ruddle in a little container,

5. i.e. formed into small balls and fed by hand.
6. Since a ewe's pregnancy lasts five months, breeding in April would produce lambs in September.

and watch the springs or nearby waters; then touch the backs of the bees as they drink with a straw dipped in that liquid, and wait there. If the ones we coloured return quickly, we shall know their lodgings are close by; if belatedly, they are at a greater distance, which we can estimate by the lapse of time.

§3 The nearby swarms you will find easily; the following method will lead you to the distant ones. Cut one internode of cane with its joints, and make a hole in its side. Place a little honey or *defritum* in it, and lay it near a spring. When the bees gather round and crawl in after the scent, you close the opening by placing your thumb over it, and allow only one to leave. Follow her flight: she shows you the direction of their lodgings. When you cannot see her, you release another straightway and follow her. Released one after another in this way, they will bring you to the location of the swarm. §4 Some people place a very small container of honey near the water sources; when the bee coming for water tastes it, she will head to where the others are feeding, and bring them into view. As their numbers grow rapidly, by noting their direction as they fly back you will follow them to the swarm.

Collecting the swarm

§5 If the swarm is hidden in a cave, it will be driven out by smoke. On leaving, it can be frightened by the sound of brass and will suspend itself from a shrub or some part of a tree; then you can bring up a vessel and collect it. But if it is in the branch of a hollow tree, the branch can be cut above and below it with a very sharp saw, then covered with a clean cloth, and brought and placed amongst the beehives. §6 They are tracked in the morning, so the whole day is available for following them. For in the evening their work is done, and they generally do not return to water sources. The vessels in which they are collected should be rubbed with *citreago*[7] or sweet herbs, and sprinkled with a little liquid honey. If the task is done in spring and beehives scented in this way are placed around springs in places which the bees frequent, they will draw a throng of bees to them without effort, so long as they can be protected from thieves.

7. Possibly lemon balm, but cf. 1.37.2 with footnote.

Spring cleaning

§7 This month is also the time, as above (4.15.4), to clean the beehives, and to kill the moths which are particularly abundant when mallows are flowering. We shall intercept them in the following way: at dusk we shall place among the beehives a bronze vessel shaped like a *miliarium*,[8] i.e. tall and slender, and set in its base a burning light. The moths will gather there and flutter around the light, and because of the slender shape of the vessel they will inevitably be killed by the proximity of the flame.

8. *Hours*

This month's hours match the hours of September, as follows:

Hour:	1	2	3	4	5	6	7	8	9	10	11
Feet:	24	14	10	7	5	4	5	7	10	14	24

8. A cylindrical vessel (like a modern domestic water heater) used to heat water for the baths, 1.39.3, named for its similarity in shape to a milestone.

Book 6: May

Field Work

1. *Seed crops and hay*

§1 In cold and wet places we shall sow millet and Italian millet, by the method described (4.3).

Now almost all the crops are flowering, and should not be touched or cultivated. The flowering takes place as follows. Cereals and barley and single-seeded plants will flower for eight days and then swell up for 40 days, after dropping their flowers, till they reach maturity. But double-seeded plants like beans, peas and the rest of the legumes flower for 40 days and swell up at the same time.[1]

§2 This month in places that are dry, hot or beside the sea, hay should be cut – certainly before the grass becomes parched. If the hay is soaked by rain, it should not be turned until the surface has dried out.

2. *Vines: training and trimming*

§1 Now we should examine the branches produced by a new vine, and leave just a few stout ones and strengthen them with supports until they harden. A vine that has been pruned and is sprouting should not have more than two or three branches left on it, and they should be tied to prevent wind damage. I say three branches should be left because of the danger that, if you leave fewer at the outset, none at all will remain if the winds shatter them.

---

1. Single-seeded plants or monocots have one seed-leaf (cotyledon), whereas double-seeded plants or dicots have two. The monocots include all the true grains.

§2 This is a suitable month to trim the vines. The right time for trimming is when the soft twigs snap without difficulty when just pulled by the fingers of the person removing them. Trimming makes the grapes fatter, and promotes their ripening by letting in sunlight.

3. *Opening new fields*

§1 Now too fields that are rich and weedy should be first-ploughed. But if you want to open up uncultivated fields, you need to assess whether the field is dry or wet, and covered in woodland or grass, shrubs or bracken. If it is wet, it should be dried by cutting ditches all across it. Open ditches are familiar, but blind ditches are made like this: trenches are sunk diagonally across the field to a depth of three feet; then they are filled halfway with small stones or gravel, and levelled off with the earth that we dug out. §2 From their starting-point the ditches should head for one open ditch, towards which the water can run downhill. In this way the moisture is led off while the openness of the field is not lost. If there are insufficient stones, they can be covered with vine-prunings or a layer of straw or any kind of brushwood.

§3 If the field is forested, the trees should be rooted out, or left only here and there, before cultivation; if it is stony, collecting the mass of rocks into field walls will both clear the field and enclose it in the process. Rush and grass and bracken will be defeated by repeated ploughing. If you frequently sow in beans or lupines, and cut the bracken as soon as it shoots with the tip of a hook, you will eliminate it within a short time.

4. *Other field tasks*

§1 This month is a suitable time for tilling the ground around trees and vines that were ablaqueated earlier, i.e. for covering their roots.[2]

---

2. The earlier ablaqueation partially uncovered the roots to receive rain and early warmth (2.1); now the roots are covered for summer by tillage (*occare* includes breaking up clods).

Now wood should be cut for making rough timber, once it is fully covered in foliage.[3] The yardstick for cutting is that a very good labourer in thick woodland can cut the area of a *modius*,[4] but a moderate worker one-third less.

§2 Now nursery beds are dug over thoroughly, and in particularly cold and rainy places olive trees are pruned and scraped of moss.

Anyone who has sown lupine for the purpose of manuring a field should turn it in now with the plough.

5. Gardens

Now will be a suitable time to trench those areas of garden that are destined to be filled in autumn with seeds or seedlings. This is a good month for sowing parsley, as said before (5.3), coriander and melons and gourds; cardoon and radishes and rue will be set out. Leek seedlings too are transplanted, so they can be quickened by irrigation.

6. Fruit Trees

Now pomegranates begin to flower in hot places. So if you enclose a flowering branch, as Martialis says, in an earthenware vessel buried close to the tree, and bind it to a stake to stop it springing back, it produces fruit in autumn proportional to the size of the vessel.

Also this month in hot places peach trees can be shield-budded.[5] In cold places the citron is grafted now, following the system that has been described (4.10.16). Now in cold places we shall plant jujube and fig, or graft them. Cuttings of palm are also set out this month.

---

3. Wood that was destined to be sawn into dimensional lumber was felled in winter, when the sap was down (12.15). But wood that was just to be rough-hewn with the axe, i.e. stripped of bark and branches, was felled in spring, when the bark peeled off easily because of the quantity of sap beneath it (Theophrastus *History of Plants* 5.1.1–4, Pliny 16.188).
4. One-third of a juger, i.e. roughly a quarter of an acre, one-tenth of a hectare.
5. This technique is described at 7.5.

Livestock

7. *Castration of calves*

§1 Now calves should be castrated, as Mago says,[6] at a tender age; the testicles should be compressed with a split length of giant fennel, and gradually broken down and crushed. He advises that this should be done under a waning moon, in spring or autumn. Others fasten the calf to a squeeze and use two narrow *stagnum* bars[7] like forceps to grip those muscles called in Greek the cremaster[8] muscles. §2 After gripping these, they stretch and cut off the testicles with a knife, in such a way that some part of them is left adhering to the ends of their muscles. This prevents an excessive loss of blood, and does not completely feminize the calves or remove their male strength. We should not permit the procedure which many follow, of forcing them to breed right after castration. For it is certain that they will procreate, but also that they themselves will die through loss of blood. The castration wounds should be smeared with ash and litharge. §3 The castrated animal should be kept from drinking, and given little food; for the next three days he should be provided with tender tree tips and soft shoots and green fodder, with the sweet taste of dew or river-water. Also after three days his wounds should be carefully dressed with liquid pitch mixed with ash and a little oil.

§4 But the following practice provides a better method of castration. Once the calf has been bound and put down, the testicles are confined by binding the skin tightly, and with a wooden bar providing pressure at that spot, they are cut off with a red-hot axe or hatchet, or better with an iron instrument fashioned for this purpose, shaped like a sword. As the edge of the glowing iron is pressed down in

6. The reference to Mago is owed to Palladius' source, Columella 6.26.

7. Palladius' source, Columella 6.26, specifies bars of wood (*regulis ligneis*). If Palladius thought metal bars preferable, there would be no need to use a specialized metal like *stagnum* (an alloy of silver and lead) for such a routine task. Probably, then, *regulis stagneis* in Palladius' text results from a misreading of *regulis ligneis* somewhere along the line.

8. 'Hanging', because the testicles hang from them. The term is still in standard medical use.

this way next to the bar, with a single stroke it prevents a long period of pain thanks to its speed, and at the same time, since the veins and skin are cauterized, the cut is protected from bleeding copiously by the scar that is formed, so to speak, with the very wound.

8. *Sheep shearing*

§1 In temperate places, now is the time to carry out the sheep shearing. We should treat the shorn sheep with an unguent made as follows. Mix the liquid from boiled-down lupine, lees of old wine, and *amurca* in equal quantities, and when all this has been reduced to uniformity, smear it on carefully. §2 Then after three days, if the sea is nearby, let the animals be dipped in the shallows. If our pasture is inland, rainwater boiled briefly with salt in the open air will have to be used to wash the animals' limbs after the shearing and anointing. It is said that sheep treated in this way remain scab-free all year and grow a luxuriant, soft fleece.

9. *Cheese-making*

§1 This month we shall curdle milk for cheese, using pure milk with rennet from a lamb or kid, or the skin that sticks to the stomach of chicks, or flowers of wild thistle or the sap of a fig tree. All the whey must be drawn off, and so it is actually squeezed out with weights.[9] When it begins to solidify, the cheese should be set in a shaded or cool place and pressed with weights that are added gradually in accordance with the degree of solidity attained; it should be sprinkled with parched ground salt, and pressed with more force as it becomes firmer. §2 After solidifying for some days the rounds should be set out on racks, not touching each other. The place should be enclosed and protected from winds, so the cheese retains its softness and richness. The faults in cheese are if it is dry or salty or full of holes. This will happen if it is not pressed enough, or gets too much salt, or is scorched by the sun's heat.

In making fresh cheese some people crush green pine-kernels and curdle the milk after mixing it with them. §3 Some curdle milk mixed with thyme that has been crushed and sieved several times.

9. i.e. not just allowed to drain naturally out of the curds.

You can create whatever flavour you want by adding the condiment of your choice, whether pepper or any tincture whatsoever.

10. *Bees*

This month the swarms begin to grow, and rather large young bees are born at the edges of the honeycombs. Some people think they are kings, but the Greeks call them gadflies and say they should be killed, because they disturb the peace of the swarm when it is at rest. Moths are abundant now, and we should kill them in the manner I described (5.7.7).

Construction

11. *Pavements and sun-terraces*

§1 Now around the last of the month pavements are made on sun-terraces. In cold regions and where frosts occur, they are heaved up by ice and ruined. But if we want to proceed, we shall lay down two tiers of planks at right angles to each other, and spread a level layer of straw or bracken over them. We shall make a level layer of rocks of a size that fits your hand. §2 Over this we spread a one-foot layer of concrete, and ram it down thoroughly with a pounder. Then, before the concrete dries, we shall lay edge to edge two-foot bricks that have one-finger channels on all sides; the channels in the bricks, which should be matched up, are to be filled with quicklime mixed with olive oil, and these interconnected bricks must cover the entire concrete layer. For all this material, once dried, will form a single solid body and will not transmit any moisture. §3 Afterwards we pour on a six-finger layer of crushed-tile mortar, and beat it frequently with sticks, so it cannot open in chinks. Then we shall press in fairly wide mosaic squares, or any kind of marble tiles or rectangles, and nothing will spoil this structure.

12. *Brick-making*

Bricks should be made this month, from white earth or clay or *terra rossa*. For bricks made in summer, through the swift action of the heat, dry out on the surface while moisture remains within; this causes them to open in cracks. They are made this way: earth that has been sifted carefully and cleaned of all coarseness should be mixed with straw, soaked for a long time and pressed into a brick-shaped mould. Then it should be left to dry, and turned at intervals to face the sun. Bricks should be two feet in length, one foot in width, four inches in depth.

Miscellaneous

Flavourings

13. Rose-flavoured wine
Steep five *librae* of roses, cleaned the previous day, in ten *sextarii* of old wine; after 30 days add ten *librae* of skimmed honey, and use.

14. Lily-flavoured oil
Infuse ten lily flowers in each *libra* of olive oil, and place the glass vessel in the open air for 40 days.

15. Rose-flavoured oil
You place an ounce of cleaned roses in each *libra* of olive oil, and suspend it for seven days in the sunlight and moonlight.

16. Rose honey
With each *sextarius* of rose juice you mix a *libra* of honey, and suspend it in the sun for 40 days.

17. Preserving fresh roses
You will preserve roses that have not yet opened, if you enclose them in a standing green cane stem that has been split, allow the fissure to close, and then cut away the cane when you want to have the fresh roses. Some people insert them in an untreated[10] pot and bury them outdoors, well covered, for future use.

10. i.e. not waterproofed with pitch.

18. *Hours*

In the measurement of hours, May corresponds to August:

Hour:	1	2	3	4	5	6	7	8	9	10	11
Feet:	23	13	9	6	4	3	4	6	9	13	23

Book 7: June

Field Work

1. *Preparing a threshing-floor*

This month an area should be prepared for threshing. First the earth should be raked clean; then it should be levelled, after being lightly broken and mixed with chaff and unsalted *amurca* (this protects the grain from mice and ants). Next it should be compressed with a cylindrical stone or some fragment of a column, which can be rolled to solidify the surface; then dried by the sun. Some people sprinkle water on the floor after sweeping it clean, and have small livestock walk about there and tread it; when the earth has been packed hard by their hooves, they look to see a solid dried-out condition.[1]

2. *Grain harvests*

§1 The barley harvest comes first, and starts now. It must be completed before the ears split and the grains fall out, since they are not clothed in little husks like wheat. Five *modii* of a full field can be cut in a day by an experienced harvester, three by a middling one, even less by an inferior one.[2] But we should let the cut barley lie awhile in the field, because it is thought to swell up in this way.

§2 Now too the wheat harvest is cut, at the end of the month, in maritime locations and in those that are warmer and dry. You know it is ready if the whole host of ears has ripened to a reddish colour and takes on a uniform golden hue.

1. The previous discussion of the threshing-floor at 1.36 is primarily concerned with its location and surroundings; this one, appropriately, with its preparation just before harvest. The element of duplication in these discussions reflects that in Columella (1.6.23–24 and 2.19–20).
2. From Columella 2.12.2 it seems that 'five *modii*' here is short for 'the crop produced by five *modii* of seed.'

A reaping machine

The areas of France that have level terrain use the following short-cut in harvesting; it saves human labour, and a day's work by a single ox does away with the lengthy time required for the entire harvest. A vehicle is built for this purpose, borne on two small wheels. §3 Its squared surface is fenced with boards, designed to slope outwards and give a larger space on top. At the front of the cart the height of the boards is less. Here a large number of small teeth are set in a row, spaced according to the size of the ears, and curving back on themselves towards the top. At the back of the same vehicle are two very short yoke-beams, shaped like the poles of a litter. There an ox is fastened with yoke and chains, his head towards the vehicle – a calm animal, of course, who will not exceed the pace set by his driver. §4 When he begins to push the vehicle through the crop, all the ears are caught in the little teeth and piled into the cart, while the snapped stalks are left behind; the height or lowness is controlled at frequent intervals by the oxherd who follows behind. And so with a few passes back and forth the whole harvest is completed in a short space of hours. This device is useful where there are plains or level places, and where the straw is not needed for use.[3]

3. Other fieldwork; harvesting pulse

§1 Now in very cold places we shall do what was left undone in May. We shall give the fields a uniform first-ploughing[4] in those areas that are weedy and chilly. We shall harrow the vineyards, dig over the nursery beds, gather vetch, cut fenugreek for fodder.

This month in cold places the harvest of pulse should be carried out. We shall find that lentils, collected and mixed with ash, keep well in oil-jars or salted fish jars, filled and immediately sealed with plaster. §2 Now too beans will be picked under a waning moon, before dawn of course; and before the moon begins to wax, they should be

3. See Figure 6 for a reconstruction of this machine. After Roman times, reaping machines were not reintroduced into farming until the mid-C19. Palladius' description and the monumental evidence for Roman reaping machines are discussed by White 1967: 157–73.
4. On the importance of first-ploughing uniformly, i.e. not leaving unturned baulks between the furrows, see 2.3.

threshed, cooled and stored. In this way they will not suffer from damaging weevils. This month lupine is gathered and, if desired, it can be sown straightway after threshing. But in the granary it must be placed well away from moisture. For in this way it keeps longest, particularly if smoke repeatedly blows onto the storage areas.

4. Gardens

This month around the solstice we shall sow cabbage, to be planted out at the beginning of August in a place that either is irrigated or moistened by the beginning of the rains. We shall also be able to sow parsley successfully, beets and radishes and lettuces and coriander, if we water them.

5. Fruit Trees

§1 This month too a pomegranate branch, as we said above (4.10.5 and 6.6), can be enclosed in an earthenware vessel, so as to produce fruit of corresponding size. Now pears or apples, where there are many fruits crowding the branches, should be thinned of the faulty ones, so that the sap that would be wasted on these can be diverted to the better fruit. This month too we shall be able to start jujube in cold places. §2 Now fig trees should be 'caprificated', as recounted under their care (4.10.28). Some people also graft them this month. In cold places peach trees are bud-grafted, and palm trees are dug around.

Shield-budding
The grafting of fruit trees called *emplastratio* (shield-budding) is practiced this month or in July. It only suits those trees that have rich sap in their bark, such as figs and olives and such, and (as Martialis says) peach.[5] §3 It is done as follows. From the shiny young productive branches you will choose a bud that has a good appearance and is definitely going to sprout. You will mark a two-finger square around

---

5. This restriction applies only to the donor tree, since Palladius says fig can be bud-grafted onto black mulberry or plane (4.10.32).

it, so it is positioned in the centre, and then delicately lift the bark with a very sharp paring-knife, to avoid damaging the bud. In the same way a 'shield' with a bud will be removed from the tree onto which we want to make the graft, in a shiny, fruitful place. §4 Then the new shield is fastened there conformably, and by the pressure of bindings around the bud it is forced to adhere without damage to the shoot; the result is that the area occupied by the former bud is closed up by attachment of the replacement bud. Then you will smear the surface with mud, leaving the bud free. You will cut away the higher branches and shoots of the tree, and when 21 days have passed you will unfasten the withy bindings, and find that the bud from an external source has miraculously joined the limbs of the alien tree.

Livestock

6. This month too is right for castrating calves, as described earlier (6.7). Now is also a suitable time for us to make cheese, and to shear sheep in cold regions.

7. *Bees*

Harvesting honey
§1 This month the beehives will be 'gelded'.[6] There will be many signs to inform us that the hives are ready for the honey to be harvested. First, if they are full, we hear the bees' murmur as muted. For empty areas in the combs, like hollow buildings, magnify sounds that strike them. Accordingly, when the murmur sounds loud and harsh, we recognize that the frames of comb are not suitable for harvesting. Again, when they hustle the drones (which are the bigger bees) out of their haunts with great determination, they attest to the honey being ready. §2 The hives will be 'gelded' in the morning hours, when the bees are sluggish and not irritated by the heat. We apply smoke from galbanum or dry cattle manure, the smoke being generated for convenience from coals placed in an earthenware pot. This vessel should be shaped in such a way that it can emit the smoke from a

---

6. A term for collecting honey from the hives.

narrow opening like that of an inverted funnel. So, as the bees leave, the honey will be cut away. To provide food for the swarm, a fifth of the comb should be left at this time. Of course mouldy and faulty combs should be removed from the hives.

Making honey and wax

§3 Now we make the honey, by collecting the combs in a very clean linen cloth and carefully squeezing them. But before we press them, we shall cut away the parts of the combs that are spoilt or contain the young, since their bad flavour spoils the honey. The fresh honey should be kept for a few days in open containers and its surface skimmed, until it ceases fermenting once the heat of the must has cooled. The finest honey will be that which flows of its own accord, so to speak, before the second pressing.

§4 This month too we make wax. After the remains of the combs have been cut up fine, the wax will be softened in a bronze vessel full of boiling water; then it will be melted in other vessels without water, and doled out into moulds.

Swarming

New swarms break away now, at the end of the month. The bee-keeper will have to be intent, because the new bees, with wanderlust in their youthful spirits, fly off unless they are restrained. When departing they linger about their entrance-way for a day or two, and should straightway be intercepted and given new hives. §5 The keeper will keep watch continuously till the eighth or ninth hour, since usually after that time they do not readily fly away or move quarters, though some have no hesitation in emerging and departing forthwith. The sign of an impending escape is that, for two or three days beforehand, their commotion and murmuring become sharper. The person checking them should frequently place his ear to the hive; when he recognizes this sign, he must be more watchful against their departure.

§6 They regularly give the same sign when they are about to fight, too. Their fighting is checked by a sprinkling of dust or drops of honeywater. The latter has a sweet power to restore harmony among the creatures that produced it. Once the armies have been pacified in this way and hang from a branch or wherever, if they hang in a single breast-shaped pendant, you will know that either the whole group has one king, or else they are all reconciled and their harmony survives. But if the community replicates itself, hanging in two or more breasts, they are in discord, and reveal that there are as many kings as you see breast-like shapes.

§7 When you see multiple clusters of bees, you must rub your hand with the juice of balm or parsley,[7] and search for the kings. They are a little larger and more elongated than the other bees, with straighter legs and small wings, of a handsome colour, shiny, and smooth without any hair – except perhaps that they have a profuse hair-like growth on their belly (they do not use this for stinging). There are others that are dark and hairy: these should be killed, and the more handsome one left. If he frequently strays with the swarm, keep him in place by pulling off his wings. For while he remains, no bee will depart.

§8 But if no new swarms are produced, we can combine the numbers of two or three containers. Then we shall sprinkle the bees with sweetened water and keep them shut up for three days with a supply of honey as food, leaving just small air-holes in the chamber. But if a hive's numbers have been drained by some disaster and you want to restore them by adding population, you will watch the honeycombs and the extremities containing the young in other hives that are packed full, and when you find indications of a king about to materialize, you will cut him away with his new generation, and place them in your hive. §9 The sign of a future king is this: among the openings which contain the young, one is visibly larger and longer, shaped like a breast. They should be transferred at the point when they have gnawed away the coverings of their cells, are mature enough to emerge, and are struggling to push out their heads. For if you transfer them while immature, they will die.

7. *Apii* (parsley) is probably an error for Columella's *apiastri* (another name for balm); this is another error that could have occurred at any stage of copying.

If a swarm suddenly rises into the air, it should be frightened by the noise of brass or earthenware. Then it will return to the hive or hang on nearby foliage; from there it should be carried by hand or in a scoop to a new container that has been sprinkled with familiar herbs and honey. When it has quietened down in that place, it should be set among the other hives in the evening.

Miscellaneous

8. *Pavements and brick*

Also this month we shall make outdoor pavements and bricks in the way I described (6.11–12).

9. *Testing future crops*

The Greeks assert that the Egyptians test the yield of each prospective crop as follows. At this time they cultivate a small tract in an area of moist, well-worked soil, and in it they sow seeds of every kind of grain and pulse in separate sections. Then at the rising of the Dogstar, which is reckoned as 19 July at Rome, they check which sowings are scorched by the rising of that star, and which are kept unharmed. They refrain from using the former, but tend to the latter, since the dry star has forecast harm or favour to each species in the following year by its present blasting or thriving.

Recipes

10. *Camomile-flavoured oil*

For each *libra* of oil you infuse one ounce weight of camomile herb in flower – the golden centre, discarding the white petals that surround the flower – and set it in the sun for 40 days.

11. *Wild vine flowers*

We collect clusters of woodland grape at the flowering stage, free from dew, and spread them out in the sun, so no moisture remains and the dried flower is ready for sifting. Then we shall screen it with a fine sieve, so the seeds cannot pass but only the flower falls through. This we keep infused in honey, and when it has been stored for 30 days, we blend it in the same fashion as is used to blend rose-flavoured wine.[8]

12. *Barley groats*

You will take half-ripe barley, which still has some greenness about it, tie it in bundles and roast it in the oven, so it can easily be ground; take care to mix some salt with each *modius* as it is ground, and store it.

13. *Hours*

June and July match each other in the length of the hours:

Hour:	1	2	3	4	5	6	7	8	9	10	11
Feet:	22	12	8	5	3	2	3	5	8	12	22

8. i.e. it is blended with old wine; see 6.13.

Book 8: July

1. Field Work

The fields that were first-ploughed in April should be second-ploughed around the first of this month.

In temperate places the wheat harvest is completed now, as described [7.2].

It will be very expedient to clear woody fields of trees and shrubs, on a waning moon, their roots being cut and burnt.

This month trees that stand in the grain fields should be shielded once the grain is cut by mounding up earth around them, on account of the sun's excessive heat. One worker will cover 20 very large trees in a day.

Now too new vines should be dug around in the morning, and in the evening once the heat drops, and the soil broken up after dislodging the grass.

This month, or before the dog-days,[1] you will usefully root out ferns and sedge.

2. Gardens

§1 Also this month we sow onions in well-watered or cold places, and radishes and orach, if we can water it, and basil, mallows, beets, lettuces and leeks, all to be watered.

This month in a watered spot we shall sow navews and turnips in soil that is loose and crumbly, not dense. Turnips love moist places and level ground, but the navew grows better in dry, almost thin soil, sloping and gravelly. §2 The character of the place can change

1. The hottest days of the year, beginning in late July.

167

the seed of each species into the other: in one soil turnips sown for two years [2] change into navews, but in another soil navew turns into turnip. They require soil that has been worked, manured and turned over, which will help them and the grain crops that are sown there the same year.[3] Four *sextarii* of turnip seed suffice for a juger, but five of navew. If they are crowded, you should thin them, so the rest can grow strong. §3 To make the seedlings bigger, you dig up the turnips, clean them of all their leaves and cut them down to a thickness of half a finger on the stalk. Then you will sink them eight fingers apart in trenches of carefully worked soil, throw in the earth and tread it down. In this way they will grow large.

3. Fruit Trees

§1 This month too shield-budding can be practised as I described earlier (7.5). Grafts of pear or apple trees made at this time in moist places have been successful in my experience.

This month too, on late-fruiting trees that have loaded their branches with excessive produce, any defective fruit that you find should be picked off, as I said before (3.25.16 and 7.5.1), so we can direct the tree's sap to nourish the better fruits.

§2 I recall that I planted a citron tree cutting at this time in a well-watered spot in a cold region and stimulated it with daily watering; both in growth and in yield it equalled my hopes for its success.

At this season in moist places the fig can be bud-grafted and the citron tree grafted, and by the middle of the month the palm tree can be dug around.

In temperate places almonds are ready for gathering now.

4. Livestock

Breeding cattle
§1 This is a particularly good time for cows to be covered by the bulls,

2. i.e. using seed from plants grown in that soil.
3. i.e. once the turnips have been harvested (about two months after being sown).

because in this way their ten-month pregnancy can reach its term in late spring; and being full of energy after the springtime fattening, they definitely love the pleasures of mating. According to Columella, fifteen cows are sufficient for one bull, and one must guard against their being unable to conceive because of excessively fat condition.[4] If there is an abundance of fodder in the area where we are pasturing, a cow can be bred annually; but if there is a lack in this regard, they should be impregnated in alternate years, especially if the cows have regularly been used for some kind of work.

Breeding sheep

§2 This month the whitest soft-fleeced rams should be chosen and put to the ewes. One should look for whiteness not only in the body but also in the tongue: if it is mottled, the result is varied colour in the offspring. From a white ram lambs of another colour are often born, but from dark rams, according to Columella, a white lamb can never be sired.[5] We shall choose a ram that is tall and long, with a low belly covered in white wool, a very long tail, dense fleece, broad forehead, large testicles, three years old – though he can usefully breed up to eight years of age. §3 A ewe should be bred at two years old, and she is well suited to bearing up to five years old; in her seventh year she stops. Choose a big-bodied ewe with a luxuriant, very soft fleece and a big woolly belly. One must ensure that sheep are well fed with rich fodder, and pastured far away from brambles, which both reduce the wool take and cut the body.

§4 The rams should be introduced in July, so their lambs can grow strong before winter. Aristotle states that if you want more males to be born, at breeding season you should choose dry days and a northerly breeze, and pasture the flock facing that wind; if you want females sired, look for a southerly wind, direct the grazing flock towards it, and breed the ewes in this way.[6] §5 Losses from

4. Columella 6.24.3.
5. Columella 7.2.6.
6. The reference to Aristotle comes from Palladius' source for this chapter, Columella 7.3. What Aristotle actually says is that more male lambs are engendered when the wind is northerly, and that shepherds believe the gender of lambs is also influenced by the direction in which their parents faced when mating (*Generation of Animals* 766b35–767a13, cf. *History of Animals* 574a1–2).

dead or unsound ewes need to be made up with new offspring. In autumn all weakly ewes should be sold; otherwise they may perish from winter cold in their frail condition. Some people prevent the rams from mating for two months beforehand, so the postponement of pleasure can increase the flame of their sexual desire. Some allow them to mate indiscriminately, so as to have a supply of lambs in this way throughout the year.

Miscellaneous

5. *Eradicating grass*

This month, when the sun is in the house of Cancer while the sixth moon is in the sign of Capricorn, grass that is removed will not regrow at all from the roots, according to the Greeks. They also say that if hoes are made of Cyprian copper and dipped in goat's blood, and after heating in the furnace are tempered not with water but the same blood, any grass that is dug up with them is killed.

Recipes

6. *Squill wine*

§1 This month we make squill wine as follows. We dry squill from the mountains or the seaside just before the rising of the dog-stars, well out of the sun. Of this we place one *libra* weight in an *amphora* of wine, first cutting off the superfluous parts and discarding the leaves that surround the end. §2 (Some people pierce these coverings and hang them on a string, so they can be infused and sunk in the wine without being mixed with the lees; then after the space of 40 days these hanging festoons can be removed.) This kind of wine will combat the cough, purge the stomach, break up phlegm, ease ailments of the spleen, sharpen eyesight, stimulate the digestive juices.

7. *Honeywater*

At the beginning of the dog-days you take pure water from a spring the day before. In three *sextarii* of water you mix one *sextarius* of honey that has not been skimmed, and after dividing it carefully among *caroenum* vessels,[7] you will have it agitated for five hours on end by pre-pubertal boys, shaking the actual vessels. Then you will let it stand in the open for 40 days and nights.

8. *Squill vinegar*

§1 You will take the tender middle part of uncooked white squill, discarding all the hard outer parts, to a weight of one *libra* six ounces, cut it up finely, and drop it in 12 *sextarii* of very sharp vinegar. The sealed vessel you will allow to stand in the sun for 40 days. Then, after discarding the squill, you will strain the vinegar carefully and transfer it into well-pitched vessels. §2 Another vinegar is beneficial for digestion and health: you put eight drachmas of squill[8] and 30 *sextarii* of vinegar[9] in a vessel with one ounce of pepper and some mint and cassia, and use it after a while.

9. *White mustard*

You will carefully reduce white mustard seed to a powder, to the measure of a *sextarius* and a half-*sicilicus*;[10] mix it with five *librae* weight of honey, one *libra* of Spanish oil, and a *sextarius* of sharp vinegar, and use it after churning it diligently.

10. *Hours*

The hours of July and June match each other with equipoised measurements:

Hour:	1	2	3	4	5	6	7	8	9	10	11
Feet:	22	12	8	5	3	2	3	5	2	14	22

---

7. *Caroenum* was a syrup produced by boiling down grape juice, 11.18.

8. A drachma is a somewhat variable Greek unit of weight, taken over by Palladius from his source here, perhaps equivalent to 4.3 g; eight drachmas would be about 1¼ ounces.

9. 30 *sextarii* is far too much: the mistake arose from a misunderstanding of λ′ = *litra* (roughly half a pint) in Palladius' Greek source as λ′ = 30.

10. Half a *sicilicus* would be ⅛ oz, but so small and precise a measurement does not fit the context here or at 12.22.3: Palladius perhaps misinterpreted an abbreviation in his Greek source.

Book 9: August

Field Work

Ploughing

1. At the end of this month, around the 1st of September, the initial ploughing should be done on land that is level, moist and lean.

Vines

In seaside locations preparations are pushed ahead now for the vintage. This is also the month for harrowing of vineyards in very cold places.

2. At this time, if the soil in a vineyard is thin and the vines themselves rather poor, you sow three or four *modii* of lupine per juger and then harrow the ground. When it has bushed up, it is turned under; it provides the best manure for the vines, since bringing dung into vineyards is unsuitable because of spoiling the wine.

3. Now vines are trimmed in cold places,[1] whereas in hot dry places the grapes are shaded,[2] if the low growth of the vines or the availability of labour make it feasible, to prevent their being dried by the strong sun.

Other tasks

Also this month we can extirpate sedge and ferns.

4. Now the pastures should be fired, so that tall fast-growing plants are shortened to the bottom of the stem, and fresh grass can grow up more vigorously in place of the dry grass that has been burnt.

---

1. To allow the sun to reach the grapes and ripen them.
2. With straw or some other covering (Columella 11.2.61).

5. Gardens

§1 The end of this month is also the time for sowing turnip and navew in dry places, by the method described earlier (8.2).

At the end of this month radishes are sown in drier places, so as to provide service in winter. They love rich, loose soil, long dug over, such as suits turnip, and dread tufa and gravel; they relish cloudy skies. They should be sown in large beds dug to a good depth. §2 They grow better in sand. They should be sown after fresh rainfall, unless they can be irrigated. The seed should be covered straightway with a light hoe. At sowing, two *sextarii* (or four, some say) will cover a juger. Spread chaff around them rather than manure, since the latter makes them pitted. They become sweeter if you sprinkle them frequently with salt water. Female radishes are thought to be those which are less sharp-tasting and have broader leaves that are smooth and an attractive shade of green. §3 Accordingly we shall collect seeds from these. It is believed that they grow bigger if you frequently remove all the leaves, leaving only the thin stalk, and bury them in soil. If they are too sharp and you want to make them sweet, you will steep the seeds for a day and night in honey or raisin wine. It is known that radish like cabbage is hostile to vines. If they are planted together, they recoil because of their discordant nature.

This month too we shall sow carrots.

6. Fruit Trees

Orchards can still be shield-budded now. Most people graft the pear now, and, in well-watered places, the citron tree.

7. Bees

This month hornets are troublesome to the beehives, and we need to track them down and kill them. Now too we should carry out the tasks that we did not tackle in July.

Water Supply

8. *Finding water*

§1 If there is a shortage of water now, you will need to search for it and trace it. You will be able to find it as follows. In those areas where you need to search for water, you will lie flat on the ground before sunrise, face down, with your chin pressed to the soil, and look to the east. Where you see the rising air wavering with a fine mist and sprinkling the dew, so to speak, you will note the spot, using a nearby bush or tree as a landmark. For it is established that where this happens in dry places, there is water below.

But to judge its scantiness or abundance, you will assess the nature of the ground. §2 Clay will produce scanty veins of water, not of the best flavour; loose grit will produce veins that are meagre, sour, muddy and deeper underground; black soil will produce trickling flows gathered from winter rains and ground-water, small but of excellent flavour; gravels will produce modest and unreliable veins but of exceptional sweetness; coarse grit and sand and *carbunculus* will produce reliable veins with a copious abundant flow. §3 In red rock they are good and abundant, but you have to take steps to prevent the veins you find from escaping through cracks or flowing away through interstices. At the foot of mountains and in siliceous rocks they are copious, cold and healthful, but in level places they are salty, noisome, warm, unpleasant; if the flavour is excellent, you will know they have come underground from a montane source. Even in the midst of plains they will emulate the sweetness of mountain springs, if they are protected by shade-trees.

§4 The following too are signs for tracing water (we can trust them if there is no hollow, and no water regularly sitting there or flowing by): thin rushes, willow scrub, alder, chaste tree, reeds, ivy and other plants that need moisture to grow.

§5 So, in those places where you find the aforesaid signs, dig a hole three feet wide and five feet deep, and around sunset let a clean vessel of bronze or lead, wiped inside with oil, be placed there upside-down on the bottom of the hole. Then let a lattice of sticks and leaves be supported on the edges of the hole, topped with soil, to cover the whole space. §6 Next day open it up, and if condensation or drops are found inside the vase, you can be certain that water is present. Similarly if a clay vase, dry but not fired, is placed and covered in the same way; next day, if a vein of water is present, the vase will have disintegrated through absorbing moisture. Similarly if a sheepskin, placed and covered likewise, gathers so much moisture that it can be wrung out next day, it will attest to the presence of a plentiful source. §7 Similarly if a lamp full of oil is lit and likewise placed under a covering, and the following day is found extinguished although fuel remains, the place will have water. Similarly if you build a fire there, and the heated earth belches out damp, foggy fumes, you will know water is there.

So when these conditions have been found, and confirmed by the certain knowledge of signs, you will dig a well and look for the water source, or, if there are several, collect them into one. Mind, water sources should be sought particularly at the foot of the northern sides of mountains, since they are more abundant there and more serviceable.

9. *Wells*

§1 In digging wells, however, one must take precautions against the danger to the diggers, since often the earth produces sulphur, alum and pitch. Their gasses when combined give off a pestilential vapour, and once they fill the nostrils they immediately stifle the breath of life, unless a person saves himself by speedy escape. So before a descent is made to the depths, place a lighted lamp there. If it is not extinguished, you will not fear danger; but if it is extinguished, you must beware of the place, which is filled with a deadly vapour. §2 If water cannot be found elsewhere, we shall dig wells to the left and right down to water level, and from these we shall open passages inwards from each side like nostrils, so the harmful vapour can dissipate.

Once this is done, the sides of the wells should be supported by concrete walls. The well should be dug eight feet in diameter, so the walls can occupy two feet on each side. The concrete should be rammed periodically with wooden bars, and it should be made with tufa or hard rock.

§3 If the water is muddy, it is corrected by an admixture of salt. But while the well is being dug, if the earth is unstable by reason of its loose nature, or is softened by moisture, you will set vertical planks against it all around, and support them with horizontal bars, lest the diggers be trapped by a collapse.

10. *Testing water*

You will test newly-found water in the following way. You sprinkle it in a shiny bronze vessel, and if it leaves no spots, it may be judged acceptable. Again, when boiled down in a bronze vessel, if it leaves no sand or mud at the bottom, it will be usable. Again, if it is able to cook legumes quickly,[3] or if it is of pellucid colour, lacking moss and any inappropriate discolouration from impurities.

11. *Aqueducts*

§1 Water from wells on high sites can emerge on lower ground like a spring (if the nature of the valley below permits), by means of a hole bored through the ground into the bottom of the well. But when water has to be led, this is done through a concrete channel or lead pipes or wooden flumes or earthenware tubes. If it is led through a channel, the conduit must be given an unbroken surface, so the water cannot escape through cracks. Its size should accord with the quantity of water. If it is coming across level ground, the structure should gradually slope, by 1.5 feet in 60–100 feet, to give a vigorous flow. §2 If a mountain stands in the way, we shall either lead the water sideways round its flanks, or create a tunnel at the level of the water source, through which the structure can make its way. But if a valley intervenes, we shall build vertical piers or arches to the correct level for the water, or allow it to descend enclosed in lead pipes and to rise on the far slope of the valley. If the water is led in earthenware tubes, which is healthier and more advantageous, they should be made two fingers thick and narrow at one end, so that one can fit into the next to the distance of a palm's breadth. We need to smear these joints with quicklime kneaded with olive oil. §3 But before a flow of water

3. The presence of calcium or salt or acid in the water greatly lengthens the cooking time of beans (cf. 12.1.3).

is allowed into them, ashes mixed with a little liquid should be run through them, to seal any flaws in the tubes. The last choice is to use lead pipes, which make the water toxic. For as the lead corrodes, white lead is formed, which will harm the human body. A diligent person will construct little reservoirs, so that even a meagre source can furnish a good supply of water.

12. *Weights of lead pipes*

The measures of the lead pipes should be observed as follows: a 100" pipe 10' long should contain 1200 lb, an 80" 840 lb, a 50" (again 10' long) 680 lb weight, a 40" 600 lb weight, a 30" 450 lb weight, a 20" 240 lb weight, an 8" 100 lb weight.[4]

13. *Unripe-grape juice with honey*

In six *sextarii* of juice from half-ripe grapes you will need to infuse two *sextarii* of vigorously whipped honey, and let it brew in the sun's rays for 40 days.

14. *Hours*

August is matched with May by the comparable course of the sun:

Hour:	1	2	3	4	5	6	7	8	9	10	11
Feet:	23	13	9	6	4	3	4	6	9	13	23

---

4. The measurements in inches are of the width of the sheet of lead, before it is bent round into a pipe.

Book 10: September

Field Work

1. *Ploughing and manuring*

§1 This month rich land that regularly holds moisture for a long time will be given its third ploughing, though in a wet year it could be third-ploughed even earlier. Now moist, level, lean land, which was first-ploughed in August as we said (9.1), is second-ploughed and sown. Slopes with thin soil should be ploughed now for the first time and sown straightway around the equinox.

§2 Manuring of the fields should be done now. The dung will be distributed more thickly on hillsides, more thinly on level ground, under a waning moon. (If this last point is observed, it will curb the weeds.) Columella states that for one juger 24 cartloads of manure are sufficient, but 18 on level ground.[1] No more piles of manure should be spread out than can be ploughed in that day, lest the manure lose its 'juice' and become ineffective.[2] §3 In fact dung is spread at any time in the winter. If for some reason it cannot be spread seasonably, sprinkle powdery manure like seed over the fields before you do your sowing, or fling goat manure by hand, and mix it into the soil with hoes. It is not helpful to manure too much at one time, but frequently and moderately. Wet ground requires more manure, dry ground less. §4 But if you do not have an abundant supply of dung, an excellent substitute for manure is to sprinkle fine clay, i.e. argil, on gritty places, and grit on places that are clayey or too compact.

1. Columella 2.5.1.
2. In warm weather manure lying on the surface would quickly lose its nitrogen through volatization.

This both benefits the grain crops and results in very fine vines – for dung in vineyards has a habit of spoiling the flavour of the wine.

Sowing

Wheat and emmer
2. This month, in places that are poorly drained or lean or cold or shaded, wheat and emmer will be sown around the equinox while fine settled weather continues, so the roots of the crop can grow strong before winter.

Remedying salty soil
3. §1 The soil has a way of discharging a salty fluid that kills the crops. Where this happens one should scatter pigeon dung or cypress leaves, and then plough to mix them in. It will be better than any remedy, however, if the noxious fluid is drawn off by means of a ditch.

Quantities of seed
On a juger of moderately good land we shall sow five *modii* of wheat and the same of emmer; rich land should receive four.

Measures against pests
If you wrap the *modius*-measure used for sowing in a hyena's skin and let the seed stay there for some time, people vouch that it will do well once sown. §2 Again, since there are subterranean creatures that frequently cut the roots and kill the standing grain, it will avail against this if the seeds to be sown are steeped for one night in the juice of the plant called *sedum*, mixed with water, or if the juice squeezed from a wild cucumber, together with the crushed root, is diluted with water, and the seed is soaked in this liquid. Some folk, when they notice their crops have this trouble, at the very start of the disorder douse the furrows and ploughland with unsalted *amurca* or with the aforesaid liquid.

Canterinum barley[3]

4. Now in thin soil *canterinum* barley is sown at five *modii* per juger. After this crop you will let the fields lie fallow, unless you scatter manure on them.

Lupine

5. Now, or somewhat earlier, lupine is sown, in any kind of ground, even in unploughed soil. It will benefit from being sown before cold weather begins. It does not grow on muddy land, abhors clay, loves thin soil and *terra rossa*. A juger is covered by 10 *modii* of seed.

Peas

6. At the end of this month we shall sow peas, in light loose soil in a warm situation; they revel in a wet climate. It will be sufficient to scatter three or four *modii* to a juger.

Sesame and alfalfa

7. Now sesame will be sown, in crumbly earth or rich sand, or in a mixed soil.[4] It will be sufficient to sow four to six *sextarii* per juger. At the end of this month we shall do the first-ploughing of fields that are to hold alfalfa.

Vetch, fenugreek and mixed fodder

8. Now is the first sowing of vetch and fenugreek, when they are sown as fodder.[5] Seven *modii* of vetch will fill a juger, and similarly for fenugreek seed. Mixed fodder too is sown, in perennial cropland which has been manured, using *canterinum* barley.[6] We sow 10 *modii* per juger around the equinox, so it can grow strong before winter. If you wish to graze it frequently, it will hold up as pasture until May; but if you want to gather seed from it as well, graze it till the first of March, and then keep the animals off it.

3. 'Horse-barley' (from the colloquial *canterius*, a nag), used to feed both farm animals and humans according to Columella 2. 9.14.
4. *Congesticia*, a mixture of soils brought together from different sources.
5. Seed crops of vetch and fenugreek were sown in January (2.6–7).
6. The barley was sown in a mixture with vetch and legumes, to produce green feed for the livestock.

Lupine as manure

9. To fertilize lean areas, lupine should be sown around the middle of this month. When it has grown, it should be cut and turned in with the ploughshare, and allowed to rot.

10. *Pasture land*

§1 Now we can lay down new pastures, if desired. If there is a possibility of choice, we shall earmark an area that is rich, dewy, level, and slightly sloping, or else valley land where moisture is not forced to run off at once, nor to remain for a long time. §2 (True, a pasture regime can be laid down even on loose thin soil, if it is irrigated.) So the area should be cleared of tree-roots at this time, and freed of all obstructions, vigorous substantial weeds, and shrubs. Then, when it has been worked frequently and broken up by much ploughing, and after the stones have been removed and clods broken up throughout, it should be manured with fresh dung under a waxing moon. §3 It should be kept strictly untouched by the hooves of draft animals, especially when it is moist, lest the pressure of their feet make the ground bumpy in soft areas.

If moss has spread over old pastures, it should be scraped off, the areas raked, and hayseed sown over them; ash should frequently be scattered there as a measure to kill the moss. §4 But if the area has become infertile through decay, neglect or age, it should be ploughed up and levelled anew, for it will generally be conducive to plough infertile pastures.

In the pasture-to-be we can sow turnips. When their harvest is finished we shall carry out the other steps mentioned.[7] After these steps we shall sow vetch mixed with hayseed. But it should not be irrigated until it has stabilized the soil; otherwise the force of the water flowing through will damage the soft surface.

11. *Making the vintage*

§1 This month the vintage should be made in places that are warm or coastal, and prepared for in cold places. In treating the jars with

7. i.e. clearing large weeds, ploughing and manuring.

pitch, the yardstick will be that a jar of 200 *congii* should be treated with 12 *librae*, and from there you subtract in proportion to the size of a smaller one. We recognize that the vintage is ripe if, when a grape is squeezed, the waste matter concealed in it, i.e. the seeds, are dark, some of them almost black – a result of natural ripeness. §2 More diligent folk blend 20 *librae* of pitch with a *libra* of pure wax, which improves the bouquet and flavour, and softens the pitch by its pliability, preventing it from splitting in cold weather. Nevertheless the pitch should be tasted for sweetness, since wine is often spoilt by its bitterness.

12. *Other fieldwork*

Now in some places millet and Italian millet will be harvested. This is the time when cowpea should be sown for eating.

Now owl snares on poles and other equipment for bird-catching should be prepared, for use around the first of October.

13. Gardens

§1 Now *poppies* are sown in places that are dry and hot. They can also be sown together with other pot-herbs. They are said to grow more successfully in a spot where twigs and prunings have been burnt.

At this time you will sow *brassicas* quite successfully, so as to plant out the seedlings at the start of November; cabbage could be produced from them in winter, and greens in spring.[8]

§2 This month you will have to trench three feet deep the areas of garden that you intend to fill with seed in the spring, and bring manure onto them on a waning moon.

At the end of this month we shall sow *thyme*. However, it grows better from divisions, though it can grow from seed too. It loves sunny, thin, coastal ground.

Now around the equinox you will sow *oregano*. It welcomes being manured and watered until strongly grown. It loves rough, rocky places.

During these same days *caper* is sown. It spreads widely, and its sap is harmful to the ground. So it should be sown, to prevent its

8. Compare 5.3.1.

advancing too far, surrounded by a ditch or walls hardened from mud, in dry thin soil. It harries weeds without help. It flowers in summer; near the setting of the Pleiades[9] caper dries up.

§3 *Git*[10] is usefully sown at the end of this month. This month we shall sow *cress* and *dill* in temperate or hot areas, and *radishes* in dry areas; also *carrots* and *chervil* around the first of October plus *lettuces* and *beets* and *coriander*, and, in the first days of the month, *turnips* and *navews*.

14. Fruit Trees

§1 Around the last of September, or in February, we shall start *azaroles*[11] by shoots or kernels. Their tender infancy must be carefully nurtured. A detached slip must be taken with roots; we shall smear it with ox manure and mud. It should be set in rich well-worked soil with shells and seaweed placed underneath, and covered in soil for most of its length. §2 Others knock the seeds out of the fruit, dry them in the sun, and plant them straightway in threes in autumn in rich soil that has almost been sieved. They are said to coalesce into a single shrub. It must be helped by constant watering and digging, which scratches the soil lightly and toughens the plant's tenderness. Then after a year or a little longer a plant grown from seed is transplanted, and in this way produces sweeter fruit.

§3 At the end of January or in February a scion of azarole does wonderfully well if grafted onto quince. But it can be grafted onto all varieties of apples, onto pears, plums and thorn, preferably in the split trunk rather than the bark. It is protected from above by a wicker basket or earthenware vase, while soil well worked with dung is piled over the scion almost to its top.[12]

9. Late October or early November, depending whether the true or apparent setting is meant.
10. *Nigella sativa*, sometimes called black cumin or Roman coriander; the blackness is that of the seeds.
11. *Crataegus azarolus*: common names include oriental medlar, Mediterranean medlar, Neapolitan medlar.
12. Being inserted in a split trunk, the scion is only a few inches from the ground, cf. 3.17.

Beneficial to azaroles are the things I mentioned as benefitting apples (3.25.13–16). Azaroles will keep if covered in millet, or in jars that have been pitched and sealed.

Miscellaneous

15. *Construction*

This month, too, we shall construct pavements on terraces and make bricks, in the way that I described under the month of May (6.11–12).

16. *Mulberry medicine*

You will have the juice of wild mulberry boiled down somewhat. Then you mix two parts of the juice to one of honey, and make sure to boil the mixture to the thickness of honey.[13]

17. *Storing grapes*

As grapes that we want to store, we should pick those that have no damage, and are neither hard through sourness nor flaccid through ripeness; they should have seeds that shine as the light penetrates, and a tough surface combined with an agreeable softness. If any are rotten or faulty, we should cut them out; and we should not allow any to remain whose impregnable sourness has hardened them against the blandishments of summer's heat. Then the cut stems of the grape bunches should be scorched in hot pitch, and hung up so in a dry, cold, dark place where no light breaks in.

18. *Grape rot*

A vine whose fruit rots through moisture must be leaf-trimmed on the sides 30 days before the vintage; one should just keep the foliage on top, which protects the crown of the vine from excessive sun.

13. This medicine is useful against rheums, spreading ulcers and inflamed tonsils according to Dioscorides 1.126. Pliny 23.136 recommends it for ailments of the mouth, throat and gullet.

19. *Hours*

The days of September and April have comparable hours:

Hour:	1	2	3	4	5	6	7	8	9	10	11
Feet:	24	14	10	7	5	4	5	7	10	14	24

Book 11: October

Field Work

1. *Sowing*

§1 At this time we shall sow emmer and wheat. The right time for sowing is from 23rd October to 8th December in temperate regions.

Now too manure is taken out and scattered.

§2 Also this month we shall sow the barley called *canterinum*. It is sown in ground that is lean and dry, or else quite rich. The latter is more than a match for it, while it cannot do any additional harm to the other sort, which could not bear any other kind of seed. In fertile land, now is the time to sow it.[1]

§3 Now too we shall sow bitter vetch, lupine, peas and sesame, as I said (10.5–9) – sesame up to October 15th – and cowpea, this last in rich soil or land that is tilled annually, using four *modii* to cover a juger.

2. *Flax-seed*

This month we shall sow flax-seed, if desired. In view of its injuriousness it is better not sown, for it exhausts the soil's fertility. But if you wish, it will be sown in a place that is very rich and moderately moist, at eight *modii* to the juger. Some sow it thickly in poor soil: in this way they ensure that the flax grows fine-stemmed.

---

1. But in September in thin soil, 10.4; see footnote there on *canterinum*. Rich soil is 'more than a match for it' in the sense of not being materially depleted by it.

Vines

3. *Records of fertility*

Now is the season for the vintage, during which the fruitfulness of the vines should be noted and recorded by marks of some kind, so we can choose between them for planting cuttings. Columella states that their fertility cannot be determined in one year but in four; in that period the true excellence of the plants is ascertained.[2]

4. *Planting and other tasks*

At the end of this month, where the climate is hot and dry, on lean dry level ground or a hillside that is steep or thin, vines are most profitably planted; I discussed this sufficiently under February (3.9). In places that are dry, hot, thin, lean, sandy, or sunny, now would be a more suitable time for the tasks discussed earlier,[3] of trenching, planting vines, pruning, propagating and reviving them, and making treed vineyards, so the glebe may be helped against its thinness by the winter rains. In this way the rains bring moisture to thirsty fields, but without damaging them when they are cut up or buried in ice, since in such places the real force of frost is unknown.

5. *Ablaqueating new vines*

§1 After October 15th all new vines, whether in dug-over ground or in holes or trenches, should be ablaqueated, so the superfluous roots produced in summer may be cut away. If these grow strong, they cause the lower roots to die, and in this way the vine will remain perched on the surface; this will make it vulnerable to cold and heat. But these rootlets should not be cut right back to the trunk, lest more grow from that spot, or else the new wound inflicted on the body of the vine be galled by the force of the subsequent cold. §2 No, we shall leave a one-finger stub[4] when we cut them back. If

---

2. Columella 3.6.4.
3. Trenching was discussed under January (2.10); the other tasks under February (3.9–16).
4. The reference of 'finger', as usual in measurement, is to its breadth (conventionally $\frac{1}{16}$ of a Roman foot), not length.

the winters are mild there, we leave the vines exposed;[5] if violent, we cover them before mid-December; if very cold, we shall spread some pigeon droppings at the start of winter around the footing of the vines. Columella says this should be done for a full five years against excessive cold.[6]

6. Propagation

In the places mentioned,[7] propagation at this time is better by reason of the fact that the vine concentrates on strengthening its roots, since it is not concerned to put forth shoots.

7. Grafting

This month some people make a practice of grafting vines or trees in very hot places.

8. Olives

§1 Now too in hot and sunny places we shall establish olive orchards in the manner and system that *February* explained (3.18). We shall also make olive nurseries in such places at this time, and do everything pertaining to the olive tree. We shall also preserve white olives as will be related later (12.22).

At this time olive trees should be ablaqueated in regions that are mild and drier, so that moisture can be guided down to them from higher ground.[8] §2 Columella advised that all suckers should be torn off;[9] I prefer that a few strong ones should always be left, so that in their mother's old age a selected one can take her place. Alternatively one that is better fed by virtue of earth being piled round it, and has developed its own roots, can be transplanted as a young tree to help create an olive orchard without the care required for a nursery bed. At this time olive orchards should be given manure, if it is available,

---

5. i.e. their roots, uncovered by ablaqueation.
6. What Columella actually recommends is ablaqueation annually for the first five years, though he does also advise adding pigeon dung against cold when re-covering the roots (4.8.3–4).
7. i.e. in 11.4.
8. By digging small drainage channels.
9. Columella 5.9.13.

at three-year intervals, particularly in cold places. Six *librae* of goat manure or one *modius* of ash will suffice for one tree. Moss should always be scraped from the trees, and they should be pruned – when their eighth year is past, according to Columella;[10] but it seems to me that every year parts that are dry, unfruitful, or somehow defective in their growth should be cut away.

§3 If a healthy tree does not bear fruit, it should be bored with a Gallic auger, with the hole drilled right to the pithy centre, and an unshaped[11] cutting of wild olive should be vigorously forced into it; when the tree has been ablaqueated, unsalted *amurca* or aged urine should be poured on. For by this coupling, so to speak, sterile trees are made fruitful. If their defect persists, you will have to graft them.

This month we shall clean ditches and channels.

9. *Rain on the grapes*

The Greeks instruct that if an excessive downpour soaks the grapes, their must, after its initial fermentation, should be transferred to other containers. Thus the water, because of its natural heaviness, will remain and settle at the bottom while the transferred wine is kept pure, separated from what had been introduced by the downpour.[12]

10. *Making oil*

§1 We shall make green olive oil now, in the following way. You collect the olives while they are speckled, as fresh as possible, and if you collect them over several days, you spread them out, so as not to heat up. Any that are rotten or dry you remove. Once you have reached the right quantity for the production area,[13] you add salt, ground or not (the latter is better), at three *modii* of salt to ten of olives, and give them a first grinding. And you will leave them salted like this

10. Columella 5.9.15.
11. Palladius' manuscript or manuscripts of Columella seem to have had both the correct reading *in foramen*, 'into the hole', which he echoes with *foramine*, and the corrupt reading *informem*, 'shapeless'.
12. When grape juice begins to ferment, its specific gravity becomes less than that of water, as its sugar is converted into alcohol.
13. The *factorium* or pressing room is described at 1.20.

in new baskets, to spend the night in the salt and take up its flavour, and in the morning they should start to be pressed; they will yield a better flow of oil since the flavour of the salt has been absorbed.

§2 Of course you will first wash the channels and all the receptacles for the oil with hot water, so they will not hold any rancid remains from the previous year. You will also not build fires too close, lest the smoke taint the flavour of the oil.

Now at the end of the month in dry hot places we shall collect laurel berries for making oil.

11. Gardens

§1 In October *endives* should be sown, to help out in winter. They love moisture and loose soil; they grow best in sand and in salty seaside places. A fairly level area should be prepared for them, so their roots are not exposed by the soil slipping off. When four-leaved they should be transferred to a place that has been manured.

Now seedlings of *cardoon* are set out. When we set them out, we cut back the tips of the roots with a knife and dip them in dung. We space them at three-foot intervals for growth, setting two or three together in a hole one foot deep. Often on dry days as winter approaches we shall mix in ash and dung.

§2 This month we shall sow *white mustard*. It loves ploughed soil or, if possible, mixed soil, though it grows anywhere. It must be hoed repeatedly, so as to be sprinkled with dust which fosters it. It is not too keen on moisture. You will leave in place the plants from which you intend to gather seed; those which you are preparing for food, you will make more vigorous by transplanting. In white mustard old seed is worthless, whether for sowing or use. When cracked between your teeth, if it looks green inside it is fresh; if white, it reveals its age.

§3 This month *mallow* is to be sown; it will be restrained from lengthy growth by winter's onset. It relishes a rich moist location, and enjoys being manured. Its seedlings are transplanted when they start to have four leaves or five. A youthful seedling takes better; a larger one wilts if moved. But their flavour is better if *not* transplanted. To prevent them shooting up into stalk, you will place little clods or pebbles at their centre. They should be set out well-spaced; they relish a constant hoe. They need to be freed of weeds without feeling movement in the root. If you make a knot in the root when transplanting the seedlings, they will become squat.

§4 Now too we shall sow *dill* in temperate and in hot locales. *Onions* are sown in this as in other months, also *mint* and *carrots*, *thyme* and *oregano* and *capers* at the start of the month. Likewise we shall sow *beets* in drier places; we shall also sow *wild radishes*, or move them to cultivated land to improve them; for this is the uncultivated form of radish. §5 We must now transplant *leeks* that were sown in springtime, to make the heads grow. Naturally they should be cultivated constantly with hoes, and the leek plants should be grasped as if with tongs and raised, so the empty space beneath the roots is filled with growth of head. *Basil* too we shall sow even now; it is said to germinate faster at this time, if it is moistened and sprinkled lightly with a drizzle of vinegar.

12. Fruit Trees

Date palm

§1 A person who wants to make provision for posterity should think about planting palm trees. This month, then, he will need to bury fresh kernels, taken not from old dates but plump new ones, and mix ash into the ground. If he wishes to use a cutting, it should be planted in April or May. It enjoys sunny, hot places. It needs fostering with water in order to grow. Loose soil or grit is what it requires, but with the proviso that when the young plant is set out, rich earth should be poured around it or beneath it. It should be shifted when one or two years old, in June or early July. §2 It should be dug around constantly, and watered to survive the unremitting summer heat. Palms are helped to some extent by salt water; their water should be infused with salt if not salty by nature. If the tree is sickly, dregs of old wine should be poured round it once it has been ablaqueated, or superfluous root hairs should be cut away, or a wedge of willow wood should be pressed into the pierced roots. It is agreed, however, that a place where palms grow spontaneously is useful for virtually no other fruits.

Pistachio

§3 Pistachios are propagated in autumn, in October, by both shoots and nuts. The better way is for the pistachio nuts to be planted in pairs, male and female. 'Male' is applied to those that appear to have long testicles concealed beneath their skin, looking as though they were made of bone. A person who wants to plant more carefully will prepare pots with holes at the bottom, filled with manured earth, and plant three pistachios in each, so that some sort of seedling will grow from all of them; the plant can be more easily transplanted from here, once it has grown strong, in the month of February. It loves a hot place but a moist one, and enjoys irrigation and sun. It is grafted on terebinth in February or March. Others have stated that it can also be grafted on almond.

Cherry

§4 The cherry loves a cold climate, and soil in a moist situation. In warm regions it stays small; it cannot sustain heat. It enjoys a mountainous region, or one in the hills. We should transplant a young wild cherry in October or November, and graft it at the beginning of January, once it has taken. Cherry seedlings can be raised if the fruits are scattered in the aforesaid months, since they will germinate extremely easily. §5 From my experience of this tree's easy nature, I can affirm that stakes of cherry wood, placed in the vineyard as supports, have shot up into trees. It can also be sown in January.

It is grafted better in November, or, if necessary, at the end of January. Others have stated that October too is a time for grafting. Martialis instructs that it should be grafted in the trunk; for me, between the bark and the wood always turns out successfully. Those who graft in the trunk, as Martialis says, will have to remove all the woolly drill-dust round the graft; he indicates that it harms the graft if it remains. §6 In cherries and all gummy trees one must be sure to graft when they have no gum, or when it has stopped flowing out. Cherry is grafted on cherry, on plum, on plane, according to others on poplar.

It loves to have a deep hole, rather generous spacing, constant digging. Diseased or dry stuff will have to be pruned, or growth that is too closely crowded, to open up the tree. It does not like dung, and is harmed by it.

§7 Martialis says cherries can be grown without stones as follows. You will cut a young tree down to two feet and split it right down to the root, carefully scrape out the pith of each half with a knife and immediately tie the two halves together with a band and smear dung on the top and the slits at the sides. After a year the scab that forms is healed over. You graft this tree with scions that have not yet born fruit, and, as he asserts, they will produce fruit without stones.[14]

If a cherry tree moulders from taking up moisture, a hole should be made in its trunk to drain it. §8 If it suffers from ants, you will have to pour on purslane juice mixed with half the quantity of vinegar, or smear the trunk of the tree in flower with wine dregs. If it is wearied by the heat of the dog-days, we shall have water collected from three springs, one *sextarius* from each, and poured after sunset on the tree's roots, without letting the moonlight on the remedy. Or we shall twine stems of stinking nightshade into a wreath around the tree trunk, or else make a bed of it close to the bottom of the trunk.

The only way of preserving cherries is to dry them in the sun till they wrinkle.

Other fruits

§9 In October some people plant apple trees in hot dry locales, and quinces at month-end; they plant service-berries and almonds in nursery beds, and scatter pine seeds. This month, or as they come ripe, fruits should be preserved and dried by the method covered in the section on each.

13. Bees

This month too the hives will be 'gelded' in the manner described (7.7). One should inspect them, and, if there is an abundance, take it; if a middling amount, leave half for winter's dearth; but if a lack of production is evident in the combs, remove none at all. The processing of the honey and wax was explained above (7.7).

---

14. At 3.29 Palladius retails from *Geoponika* 4.7 a similar system for producing seedless grapes.

Wine

Seasoning and care: advice from the Greeks and others

Categories of wine

14. §1 So as not to omit what I have gathered in reading, I have taken care to expound what the Greeks on their own authority have advanced as methods of seasoning wine. They differentiate the nature of wine in this way, and find this distinction: what is sweet, they call rather heavy; what is white and somewhat salty, they say is good for the bladder; what has an attractive yellow colour, is well suited to the digestion; what is white and astringent, is helpful for loose bowels; wine from overseas induces pallor and does not produce so much blood; from black grapes comes robust wine, from red ones sweet wine, but from white ones generally a middling kind.

Seasoning

§2 In seasoning wine, then, some of the Greeks add must, boiled down to a half or a third, to the wine. Other Greeks give the following instructions. Clean seawater from a clear calm sea is put in jars a year ahead and stored. Its nature is such that over this time it loses its saltiness and bitterness and odour, and becomes sweet with age. §3 So they mix an eightieth part of it with the must, and a fiftieth part of gypsum. Then after the third day they stir it vigorously, and swear that it adds not just longevity to the wine, but also brightness of colour. The wine should, however, be shaken and checked every ninth day, or at latest every eleventh. For frequent inspection will allow one to judge whether that particular kind should be sold or kept.

Some authorities drop three ounces of powdered dry resin in a jar and stir it, and argue that the wine can be made diuretic in this way.

Boiling down watery must

§4 Must that is thin on account of frequent rains, they instruct, should be remedied in the following way, which can be tested by the taste of the must.[15] They instruct that the must should be boiled down until a twentieth of the whole amount is removed. It also becomes

---

15. Another remedy, also from the Greeks, was retailed above at 11.9.

better if you add a hundredth part of gypsum. But the Spartans, they say, boil it down until a fifth of the wine is gone, and bring it out for use in the fourth year.

Making harsh wine sweet

§5 They teach that harsh wine can be made sweet if you work two *cyathi* of barley flour into a paste with some wine, place it in the vessel of wine and leave it there for an hour. Some authorities mix in dregs of sweet wine. Some add a certain amount of dry liquorice root, and use the wine after mixing it by giving the jars a long shaking. Wine can also be given an excellent scent within a few days, they say, if you drop some dried crushed berries of wild mountain myrtle in the jar and let them stay there for 10 days; then you can strain and use the wine. You can also have the flowers of an arbustive vine[16] gathered and dried in the shade. Then, once they have been diligently crushed and sifted, keep them in a new pot; when you wish, add one measure of flower (what the Syrians call a *choenica*)[17] to three jars of wine,[18] seal the container, open it on the sixth or seventh day, and use.

§6 They teach that wine is made sweet for drinking as follows: matching amounts of fennel and savory are plunged in the wine and stirred, or else the nuts from two pine cones, baked and tied in a cloth, are placed in the vessel and sealed, and this is ready for use when five days have passed.

New wine into old

§7 Furthermore new wine is made to seem old if you roast together what you judge a sufficient quantity of bitter almond, wormwood, the gum of a fruit-bearing pine[19] and fenugreek, crush them together and add one *cyathus* of this mixture per *amphora*: you will make great wines! But if you feel that the wines are about to spoil, you will mix into this recipe equal amounts of aloe, myrrh and saffron-oil residue,

16. A vine grown in an *arbustum* or treed vineyard, where the vines are trained on growing trees.
17. In fact *choenica* is a Latin form of the pure Greek word *choinix*, a measure equivalent to two *sextarii*, i.e. roughly one litre or two pints.
18. As a measure, a jar (*cadus*) is equivalent to 1½ Roman *amphorae*, i.e. about 39 litres.
19. The species that produces edible pine nuts, *Pinus pinea*.

after crushing and reducing them to powder, along with honey, and use one *cyathus* of this mixture to season one *amphora* of wine.

§8 Also, to make one-year-old wine appear long-lived, you crush and sift one ounce of melilot, three ounces of liquorice root, the same quantity of Celtic nard, and two ounces of hepatic aloe,[20] plunge six spoonfuls[21] in 50 *sextarii* and place the vessel in the smoke.

Changing the colour of wine

§9 Dark wine can be changed to white, they say, if one adds bean meal to the wine or pours the whites of three eggs into a flagon and stirs it for a long time; next day it is found to be pale-coloured. But if meal of African pea[22] is added instead, it can change the same day.

§10 They say also that the nature of vines is such, that if a white or black one is reduced to ash and added to the wine, each imposes its own colour on the wine, so that it is turned dark by a black vine and pale by a white one. The method is that one *modius* of the burnt vine wood should be placed in a storage jar containing 10 *amphorae*, and after being left thus for three days, the jar should then be covered and sealed with mud; the wine is found to be white, or black if that was the intention, when 40 days have passed.

Making mild wine strong

§11 They also claim that mild wine is made strong as follows. Leaves or roots or tender stalks of marsh mallow, i.e. hibiscus, are boiled and placed in the wine, or gypsum or two *cotulae* of chickpea or three cypress balls[23] or a handful of box leaves, or celery seed or the ash of vine-cuttings left powdery by the force of the flames, with all solid matter removed.

Clearing wine

§12 They say sour wine is made clear and excellent the same day, if you grind together 10 pepper seeds and 20 pistachios with a little

---

20. Dioscorides 3.22 says the medicinal juice extracted from this plant is liver-coloured.
21. *Coclearia*, about the size of small teaspoons.
22. Not identified.
23. i.e. the round cones.

wine, and place it in six *sextarii* of wine after first stirring it all for a long time; then you should let it rest, and strain it when about to use it. Again, cloudy wine is rendered clear straightway if you place seven pine-nuts in a *sextarius* of wine, stir it for a long time and let it stand a little; it soon becomes pure, and should be strained and put to use.

Turning must into vintage wine

§13 Again, as the oracle of Pythian Apollo is said to have revealed to the Cretans,[24] must becomes clear and takes on the flavour of an old vintage in the following way. You take four ounces of camel grass,[25] four ounces of hepatic aloe, one ounce of best mastic, one ounce of cassia quill, half an ounce of Indian nard, one ounce of the best myrrh, one ounce of pepper, one ounce of frankincense, not rancid; you pound all these and reduce them to the finest powder, shaking them through a sieve. §14 When the must has fermented, you will skim it and throw out all the grape seeds that the fermentation has brought to the surface. Then you place three Italian *sextarii* of ground sieved gypsum in 10 *amphorae* of wine, but first you transfer one-fourth of the wine that is to be seasoned into other vessels, and then add the gypsum and stir the container vigorously for two days with a green cane that has roots. §15 On the third day you place four level spoonfuls[26] of the aforementioned powder in every 10 *amphorae* of wine, add in the one-fourth of the wine that you had decanted as indicated above, and fill the container. Again you will make sure to agitate it for a long time, so the power of the ingredients can impregnate the entire quantity of the must. §16 Then you will cover and seal it, leaving a small hole for the seething wine to breathe; but in 40 days' time you close this air-hole too, and then, when you wish, taste the wine. Above all remember to ensure that whenever the wine is stirred, a pre-pubertal boy does this, or someone sufficiently chaste. Also you should seal the container not with gypsum but with ash from vine-prunings.

24. This response of the Pythian oracle is not attested elsewhere.
25. *Cymbopogon scoenanthus* Spreng.
26. *Coclearia*, as in §8 above.

Medicinal wine

§17 Again, they give the following recipe for a wine that will be healthful against infectious disease and will aid digestion. In one *metreta* of first-rate must, before it ferments, you place eight ounces of ground wormwood wrapped in a cloth, and when 40 days have passed you make sure to remove it. You pour the wine into smaller flasks, and use it.

Seasoning with gypsum

§18 If people make a practice of treating wine with gypsum, they season it now, when the first gush of the foaming must has been discharged. If the character of the wine is mild and watery-tasting, it will suit to add two *sextarii* of gypsum to 100 *congii*. But if the wine is more substantial by nature, half of that quantity will be amply sufficient.

Speciality wines

Rose-flavoured wine without roses

15. At this time you will make rose-flavoured wine without roses as follows. You place green leaves of the citron tree in a palm basket and plunge them in a vessel of must that is not yet fermenting, and stop it up; in 40 days' time, after adding honey according to the quantity of the rose-wine, use it when you wish.[27]

Fruit wines

16. This is the month to make any and all of the fruit wines that you can read about in their own sections.[28]

Honey wine

17. §1 You take whatever quantity you want of the must from large superior vines, 20 days after it has been removed from the vat. Add in a fifth part of first-rate unskimmed honey that has first been whipped forcefully until white, and mix vigorously with a rooted cane. §2 You will stir it in this way for 40 days on end, or better

27. For rose-flavoured wine using roses, see 6.13.
28. See index under 'wine'.

50, covering it after stirring with a clean cloth, through which the fermenting contents can easily breath. So after 50 days you remove anything floating on the surface with clean hands. You carefully seal it in a vessel with gypsum and set it aside to mature. §3 But it is better if you pour the wine into smaller pitched vessels the following spring, cover the vessels carefully with gypsum, and store them in a cold underground cellar or bury them partway in river sand or in the soil. As long as you carry this out carefully, no amount of ageing will spoil this wine.

Boiling down must for syrup

18. This is the time to make *defritum*, *caroenum* and *sapa*. Though these are all made in the same way from must, the method used will change their qualities and names. For *defritum*, whose name comes from *defervendum* 'boiling down', is produced when it has seethed down vigorously to a thick consistency; *caroenum*, when one third has vanished and two thirds remain; *sapa*, when it has been reduced to one third. It is improved if quinces are cooked with it and fig wood is used in the fire.

Raisin wine

19. §1 Raisin wine will be made this month, before the grape harvest. (The whole of North Africa regularly makes rich, delicious raisin wine.) If you use it for spiced wine in place of honey, you will protect yourself from flatulence. The grapes, then, are gathered after drying[29] (the more the better) and enclosed in baskets made of rushes rather loosely woven. §2 First they are forcefully beaten with rods; then, when the pounding has broken up the grapes, a basketful is placed under the screw and pressed. What flows out of it is raisin wine, and it is kept stored in a vessel like honey.

Quince wine

20. §1 After discarding the peel you will cut ripe quinces into very small thin pieces, and throw away the hard core. Then you will boil them down in honey, until reduced to half quantity, and while

29. The grapes for raisins were dried in clusters on the stalk, either on the vine as in 22 below, or on racks as at Columella 12.39.1.

cooking sprinkle on fine-ground pepper. §2 Another method: you mix quince juice, two *sextarii*, one and a half *sextarii* of vinegar and two *sextarii* of honey, and boil it down until the whole mixture resembles pure honey in thickness. Then have two ounces each of ground pepper and ginger mixed in.

21. *Saving yeast*

You will make flour from new wheat after threshing, and have it infused with must that has been pressed out underfoot, with a *laguna* of must added to a *modius* of flour;[30] then you will dry it in the sun and infuse it again in the same way and dry it. After doing this a third time, you will form from it really small loaves of the must bread, and after drying in the sun, store them in new earthenware vessels and seal them. You will use this as yeast at whatever time of the year you want to make must bread.

22. *Raisins*

You will make Greek raisins as follows. Choosing bunches of superior grapes, sweet and bright, you will twist their stems, leaving them on the vine, and allow them to wilt naturally, then remove them and hang them in the shade. You store the dried-up grapes in vessels: you place them on layers of vine leaves that are cold and dry, and press them down by hand, and after filling the vessel you again add vine leaves that are equally cool. Then you will cover them with a lid and set them in a cold dry place safe from the damaging effects of smoke.

23. *Hours*

October has March as its equal in the length of the shadows:

Hour:	1	2	3	4	5	6	7	8	9	10	11
Feet:	25	15	11	8	6	5	6	8	11	15	25

30. The volume represented by a *laguna* ('flagon') is unknown, but you would need a considerable amount of must to moisten a *modius*, i.e. 8¾ litres, of flour.

Book 12: November

Field Work

1. *Sowing*

Wheat, emmer, barley

§1 This month we shall sow wheat and emmer; this is the regular sowing at the customary seed-time. A juger will be covered by five *modii* of either kind of seed. We can also still sow early barley at this time.

Beans

At the start of this month we broadcast beans, which need a very rich or manured locale, or a valley fertilized by moisture coming from the high ground. First they are sown, next ploughed over and then ridged. They need to be harrowed extensively, so as to be well hidden. §2 In sowing beans some say one should avoid breaking up the clods of earth in cold regions, so the sprouting plants can be sheltered and protected by them in time of frosts. Sowing this species, according to the general opinion, does not fertilize the ground, but does it relatively little harm. According to Columella a field that was empty the previous year is better for grain crops than one that produced the stalks of a bean crop.[1] A rich juger is covered by six *modii*, an ordinary one by more. §3 It grows well in dense soil, but does not tolerate a lean or foggy locale. One should take special care to sow it on the fifteenth day of the moon, as long as the moon has not yet felt the force of the sun's rays. Others say the preferable choice is the

---

1. Columella 2.10.7.

fourteenth.[2] The Greeks assert that bean seeds soaked in a capon's blood are not harmed by pernicious weeds; that if steeped in water the previous day they germinate faster; and that if sprinkled with water mixed with *nitrum* they are not resistant to cooking.[3]

Lentils, linseed
Now the first lentils are sown, in the manner related under February (3.4). Also linseed can be sown throughout this month.

2. Other tasks

Particularly at the start of this month we can establish new pastures in the way described above (10.10).

Also throughout this month, in places that are hot and dry or sunny, the planting of vines will need to be performed. Now too is the right time for their propagation, and in cold places it will be timely to dig around new vines and young trees and cover them.[4] And before the middle of this month any 'diver' (i.e. curved layer) that was pinioned three years ago will be detached from the vine.[5]

3. Reviving an old vine

Now, or from this time on, an old vine on a frame or pergola, if its trunk is firm and sound, should be ablaqueated and well fed with dung and pruned quite closely. Then it should be punctured with the sharp point of an iron tool between three and four feet from the ground in the greenest part of the bark, and stimulated by frequent digging. For, as Columella states,[6] it will usually produce a bud at that point, and in spring it will put forth growth which serves to resuscitate the ancient vine.

2. One wants the moon to be full, to encourage the fullness of the crops, but not to have begun waning. It was believed to start waning when it became visible on the horizon at sunrise (hence 'felt the sun's rays'). So the fourteenth day of the moon might be safer, to avoid any possibility of waning influence.
3. The *nitrum*, which was alkaline, might be thought to counteract impurities in the cooking water that would slow the beans' cooking time; cf. 9.10.
4. i.e. cover the roots with soil, to protect them from cold; cf. 11.5.
5. Cf. 3.16.2 with footnote.
6. Columella 4.22.3–4.

4. *Pruning*

§1 Now the autumn pruning of vines and trees is performed, especially where we are encouraged by the mild climate of the region; olive orchards are pruned too. And the olives from which the first oil will be made are collected when they start to be parti-coloured. (When they are completely dark, they compensate for their less attractive appearance by an abundant flow of oil.)

§2 Pruning olives and other trees is useful in ensuring (if the regimen of the area allows it) that, with the tree-tops cut out, the branches spread outwards and then downwards at the sides of the trees.[7] But if you are faced with an uncultivated and unfenced area, you must proceed by first clearing the tree completely of low growth, so that once it has surpassed the height that animals can reach, i.e. the level of possible damage, the tree can bend downwards, safeguarded by its own size.

5. *Olives, etc.*

Now too olive orchards are planted in hot locales and dry areas, as was discussed under February (3.18). This tree loves to be lifted somewhat above moisture by an elevated situation, to be scraped of moss repeatedly, to fatten on copious manure, and to be stirred gently by fruitful winds. This month too we shall treat sterile olive trees with the above-mentioned remedies (4.8.2).

Now is a good time for making baskets, stakes and vine-props. Now is also the proper time for making laurel oil in temperate locales.

6. Gardens

This is a good month for seeding garlic and *ulpicum*, preferably in ground that is light-coloured, after digging and breaking it up without manuring. You will make furrows in the beds and place the seeds on the ridges, four fingers apart and not pressed in too deeply. You will hoe frequently, they will grow more that way. If you want to make them large-headed, tread the stalk down when it starts to appear; that

---

7. Keeping the branches low increases productivity by reducing wind damage to the blossom (*Geoponika* 9.9.9–11).

way the sap will return to the cloves. People say that if it is seeded when the moon is below the earth and again pulled up when the moon is hidden below the earth, it will have no foul odour. Garlic will keep if stored in chaff or hung up in the smoke.

Now onions too can be seeded and cardoon seedlings planted out; wild radishes too are seeded, and summer savory.

7. Fruit Trees

Peach

§1 This month in hot places – but in January elsewhere – peach stones should be placed in trenched beds two feet apart, for transplanting when the seedlings have sprung up. The stones should be set point down and not buried more than two or three palm-widths deep. Some people dry the seed-stones for a few days beforehand and store them in wicker baskets in loose soil mixed with ash. But I have often kept them till seeding time without any care. §2 Certainly they grow in any kind of locale, but they are outstanding in fruit and leaf and longevity if they are granted a hot climate and sandy, moist ground; in cold places, and particularly in windy ones, unless they are protected by some kind of barrier, they die.

While the shoots are tender, they should frequently be dug around and freed of weeds. We shall do well to transplant the sapling to a small hole when two years old. They should be set not too far apart, so as to spare each other the sun's heat. §3 Through the autumn they should be ablaqueated, and manured with their own leaves. Pruning of a peach should only involve removal of dry and mouldy twigs; for if we cut any green growth, it dries up. A wilting tree should have dregs of old wine mixed with water poured around it.

According to the Greeks a peach will grow with writing on it, if you bury its stones and after seven days, when they start to split, you open them and remove the kernels and write anything you like on them with cinnabar;[8] then you bind them up and bury them with their stones, carefully fitted together.

8. A red pigment apparently derived from a tree resin, not the modern mineral cinnabar. For a similar trick involving almonds, see 2.15.13.

Their types are these: hard-fruited, early-ripening, Armenian.

§4 If this tree becomes parched by the sun's heat, it should be covered by frequent earthing-up of soil, assisted by evening watering, and protected by being shielded from the sun. Even hanging up a snakeskin in it is helpful. Now too against the frosts a peach tree should be liberally provided with dung, or wine dregs mixed with water, or (more useful) water in which beans have been cooked. §5 If a peach tree suffers from worms, ash mixed with *amurca* will wipe them out, or ox urine blended with a third part of vinegar. If the fruit tends to fall, a wedge of mastic or terebinth wood is jammed into the trunk or a root that has been laid bare, or else a hole is drilled in the tree's centre and a willow peg inserted. If the tree makes fruit that is wrinkled or rotten, the bark should be cut away round the bottom of the trunk, and when a reasonable amount of moisture has flowed out, the wound should be sealed with clay or a mixture of mud and chaff. §6 A peach bears large fruit if, while it is flowering, you lavish three *sextarii* of goat milk on it daily for three days. Against peach ailments it is efficacious to tie on a *spartum* rope, or hang a rope sandal from the branches.[9]

Peach grafts should be made in January or February in cold places, but in November in hot places, preferably close to the ground using the fatter shoots that grow near the trunk. For scions from the top of the tree will not take hold, or not last long. It is grafted onto peach, almond, plum, but Armenian and early-maturing varieties attach better to plum, hard-fruited to almond, and attain a good age. §7 In April or May in hot places, but at the end of those months or in June in Italy, a peach tree can be bud-grafted (the process called shield-budding) by cutting the face of the trunk high up and fastening on several buds, as has been described (7.5.2–4). §8 Peaches become red if grafted onto a plane tree.

Hard-fruit peaches are kept stored in brine and honeyed vinegar, or are stoned and hung to dry in the sun like figs. Again, I have often seen hard-fruits stoned and preserved in honey, having an agreeable flavour. Again, they keep well if you fill the navel of the

---

9. *Spartum* is the name of two fibrous plants used in making rope: esparto grass and Spanish broom. Rope sandals were sometimes used to protect the feet of farm animals, cf. 14.12.2–4.

fruit with a drop of hot pitch, so as to immerse them in *sapa* in a sealed container.

Pine [10]

§9 The pine is believed to benefit everything sown beneath it. We shall seed pine by means of kernels, in October or November in hot dry regions, but in February or March in cold moist regions. It loves a locale with thin soil, often a maritime one. Amidst mountains and rocks it is found to be larger and taller; in windy moist places the trees' growth is more vigorous. But whether you want to seed mountainsides or areas of whatever kind, you will assign land to this species that could not be used for another. §10 You will plough these places thoroughly and clear them, and sow the seed like grain and take care to cover it with a light hoeing; it should not be buried deeper than a palm's breadth. The young tree needs to be protected only against cattle, lest it be trampled while fragile. You will do well to steep the kernels for three days in advance.

§11 Some say the pine fruit becomes softer through transplantation. They raise the seedlings as follows. First they plant many seeds in little cups filled with soil and dung. When these germinate, they keep the sturdier one and remove the others. When it has made suitable growth after three years, they transfer the seedling together with the cup, which they break in the planting hole to allow the roots generous space. They also mix dung in equal quantities with the soil, making alternate levels one above the other. §12 One must ensure, however, that the root, which is single and straight, is transferred whole and unharmed right to the tip.

Pruning stimulates young pine trees so much, in my experience, as to double the growth you had expected. Pine nuts can stay on the tree till now, and will be riper when gathered; but they should be gathered before they open. Unless cleaned, the kernels will not last; but some claim that they keep if placed with their shells in new earthenware pots filled with soil.

---

10. Palladius mentions the use of pine kernels in flavouring cheese (6.9) and in flavouring or clarifying wine (11.14.6, 12).

Plum

§13 If plums are seeded in autumn by means of their stones, they should be set in November, in crumbly, well-worked soil, at two palms' depth. The stones can also be planted in February, but then they should first be steeped for three days in lye to make them germinate quickly. They are also set out as shoots, which we shall take from around the trunk at the end of January or around mid-February, smearing the roots with dung. They enjoy a lush, moist locale; they thrive better in a hot climate, though they can endure a cold one too. §14 In stony or gravelly places they are helped with manure, which protects the fruit from growing wormy and liable to fall. Suckers from the roots should be removed, except the straighter ones that will be kept for transplantation.

If a plum tree is sickly, *amurca* diluted with equal parts of water should be poured on its roots, or undiluted ox urine, or aged human urine mixed with two parts of water, or ash from the oven or particularly ash from vine prunings. §15 If there is a heavy fruit drop, drill a hole in the root and drive a peg of wild olive into it. Worms and ants will be wiped out if red ochre with liquid pitch is daubed on, but lightly because of possible harm to the tree; the remedy may have the same effect as poison! Frequent watering and repeated digging are helpful.

Plum is grafted at the end of March, in the split trunk rather than in the bark, or in January before it begins to weep gum. It is self-grafted and accepts peach, almond or apple, but makes the last inferior and small. Plums are dried in the sun, laid out on racks in a fairly dry location. §16 Some people cover freshly picked plums in boiling seawater or brine, and after removing them have them dried in a moderate oven or in the sun.

Chestnut

§17 Chestnut is propagated both by volunteer plants and by seed. But that propagated from volunteers is so sickly that for two years its survival is often in doubt. So it should be propagated from the chestnuts themselves, i.e. its own seeds, in November or December, or again in February. The chestnuts chosen for planting should be fresh, big and ripe. If we plant them in November, the fruit is available to hand. §18 But if we plant in February, we must keep them till then, as follows. The chestnuts should be spread out in the shade and dried. Then they are transferred to a dry confined space and piled in a heap, and should all be covered carefully with river sand. After 30 days you remove the sand and place them in cold water. The sound ones sink; any that are damaged float. Those you have approved, you will cover again in the same way, and after 30 days you test them similarly. When you have done this thrice, you should plant those that remain unimpaired, up to the beginning of spring. Some people keep them in pots, putting sand in with them.

§19 They like soil that is soft and loose, but not sandy. They grow in grit as long as it is moist. Black soil is suitable for them, also *carbunculus* and tufa that has been thoroughly broken up. It can hardly grow in heavy farmland or *terra rossa*; in clay and gravel it cannot germinate. It loves a cold climate but will accept a mild one, if the moisture is to its liking. It enjoys sloping ground and shaded areas, particularly those facing north. §20 A place intended for this species, then, should be dug to a depth of one and a half or two feet, either entirely or in trenches marked out in a regular pattern, or at least broken down by cross-ploughing. When manured generously and reduced to fine tilth, it should have the chestnut seed planted in it no deeper than nine inches. §21 Each seed should have a twig planted next to it as a marker. The seeds should be set three to five in each spot, at four-foot intervals. Those who wish to transplant should transplant two-year-old plants. The place should have drainage channels, to prevent moisture sitting there and smothering the seed with mud. Anyone who wishes can bend the lowest shoots, which grow from the roots, into layers to propagate chestnuts. One should dig around the chestnuts in a new plantation repeatedly. In March and September it puts on more growth if helped by pruning.

§22 Chestnut can be grafted, as I have confirmed by my own experience, beneath the bark in March or April, but it responds to both methods.[11] It can also be bud-grafted. It is grafted on itself and on willow, but from willow it matures more slowly and becomes harsher in flavour.

11. i.e. under the bark or in the trunk

Chestnuts keep if they are spread out on wicker racks, or plunged in grit so as not to touch each other, or stored in new earthenware vases sunk in a fairly dry location, or enclosed in containers made of beech twigs coated with mud so you leave no airholes, or buried in very fine barley chaff, or cached in tight-woven baskets fashioned from marsh rushes.

Fruits treated in other months

§23 This month in hot dry regions we set out saplings of wild pear, so we can graft them later, and apple or pomegranate and quince and citron and medlar, fig, service tree, carob, also saplings of wild cherry to be grafted later, also cuttings of mulberry, and almond kernels and walnuts, if they are planted in nursery beds as has been described (3.10).

8. Bees

§1 At the start of this month, bees make honey from the flowers of tamarisk and other wild shrubs. This honey should not be taken, as it is being stored for winter. During this month the hives should be cleaned, as throughout the winter we cannot move them or open them. Mind, this should be done on a sunny mild day, preferably using the feathers of larger birds that have some stiffness, or something similar, to clean all the inner areas that the hand cannot reach. §2 Then we should smear all the chinks on the outside with a mixture of mud and ox dung, and pile broom or other coverings on top like a portico, to protect them from the cold and from bad weather.

Miscellaneous

9. Rank-growing vines

In hot sunny places, now will be a suitable time to prune closely those vines that lack fruit but are rank in leaf, and that make up for their poverty in bearing with an abundance of foliage. In cold places, however, they should be pruned in February. If this failing persists, we shall have to dig around them and heap up river sand or ash. Some people insert stones in the angles of the roots.

10. *Sterile vines*

The Greeks recommend that at these same times and locations a sterile vine should be treated thus: they state that the trunk should be split and a stone inserted, and that four *cotulae* of aged human urine should be poured around the trunk, so this spray can penetrate down to the roots. Then dung mixed with earth should be added, and all the soil around the roots should be turned over.

11. *Rose beds*

Although rose beds should be planted in February (3.21.1), nevertheless in seaside locales, and those that are hot and sunny, we can start a rose garden this month too. If you lack young plants but want to have the garden well supplied from a few stems, you will have to cut four-finger budded cuttings, joints and all, plant them horizontally like layers, help them with manure and watering, transplant them a foot apart when they are one year old, and in this way fill the ground that you assign to this species.

12. *Keeping grapes on the vine*

§1 On the Greeks' authority, in order to keep grapes till the start of spring on the vine, you will dig a hole in a shaded spot, three feet deep and two feet wide, round a vine that is full of fruit, and put in grit, and set stakes in it. Then you regularly twist the branches that are full of fruit and tie them onto the stakes (without damaging the grape clusters) so they do not touch the ground, and cover them so the rain cannot get in. §2 Again, on the Greeks' advice, if you want to keep grapes on a vine or fruit on a tree for an extended period, place them in earthenware pots that have holes in the base, and let them hang after carefully covering them on top – though fruit also keeps for a good time if coated in gypsum.

13. Livestock: Sheep and Goats

Lambing

§1 This month the first lambing occurs. As soon as it is born, the lamb should be moved by hand to the mother's teats. But first one should milk out a little of the milk which has a thicker character, which shepherds call *colostra*, for this will harm the lambs unless removed. The lamb should be first be penned with its mother for two days, then kept in a dark warm enclosure; with the flock of youngsters secluded in this way, the dams should be let out to pasture. §2 It will suffice, before the mothers leave in the morning and when they return full towards evening, to allow their udders to be drained by the lambs. Until they grow strong, they should be fed in the shed on bran or alfalfa or (if you have a supply) heaps of barley meal, until such time as their age and gradually increasing strength allows them to share the pasture with their mothers.

Management of grazing

§3 Pasture that springs up in newly ploughed fields or in drier meadows is suitable for sheep, but marshy grazing is noxious to them, and woodland grazing damaging to their wool. A frequent sprinkling of salt, either over the pasture or offered regularly in their troughs, should prevent the animals wearying of their feed. Through the winter, if grazing is short, hay or chaff or vetch should be provided, or more easily obtained sustenance from stored elm or beech leaves. §4 In summer months they should be pastured at dawn, when the sweetness of the tender grass is made more appealing by the admixture of dew. In the increasing warmth of the fourth hour a drink from a clean stream or well or spring should be provided. Shelter from the hot midday sun should be sought in a valley or under a shady tree. Then, when the heat diminishes in the afternoon and the ground grows moist with the first sprinkle of evening dew, we should recall the flock to pasture. §5 But in the dog days of summer, sheep should be pastured in such a way that the heads of the flock always face away from exposure to the sun. In winter and spring they should not go out to pasture until the frost melts, for frosty grass will cause disease in these animals. And then it will suffice to water them once.

Special breeds

It is customary to feed Greek sheep, like Asiatic and Tarentine sheep, in the shed rather than the field, and to lay perforated boards on the ground where they are folded, so the drainage of moisture renders their bedding safe against damage to the valuable fleece. §6 But on three days during the course of the year, in sunny weather, the sheep should be washed and rubbed with olive-oil and wine. On account of snakes, which commonly hole up under stalls, we should frequently burn cedar or galbanum or woman's hair or stag's horns.

Goats

§7 Now the he-goats should be let in, so the kids can be born in favourable conditions as early spring arrives.[12] A he-goat should be selected who seems to have two growths hanging under his jaw, big-bodied, stout-legged, with a short full neck, heavy floppy ears, a small head, and glossy, thick, long hair. He is suitable for breeding the females even before he is a yearling, but does not last beyond six years. A she-goat should be selected of similar body but with a large udder. §8 You should have fewer she-goats penned in one enclosure than you would sheep, and it should be free of mud and dung. In addition to an abundance of milk, the kids should frequently be offered ivy and branch-tips of arbutus and mastic trees. Three-year-olds are best able to raise kids; what younger mothers bear should be sold. But beyond eight years breeding females should not be kept, because these animals become barren at a greater age.

14. Collecting acorns

At this time we should think about the business of collecting and preserving acorns.[13] This task will easily be handled by female and young workers, like the collection of berries.

15. Cutting timber

§1 Now timber should be cut for building, on a waning moon. But

12. The gestation period is approximately five months.
13. As feed for livestock, or to sow new trees (cf. Palladius 3.26.3, Columella 11.2.83, 101).

you will allow the trees that are to be felled to stand for some time after they have been chopped as far as the pith, so that any moisture contained in the veins can flow out through that area. The following trees are particularly useful. The fir called Gallic Fir is light and rigid, unless rainwater leaks onto it, and lasts forever in dry buildings. Larch is very useful: if you fasten boards of it below the tiles on the front and round the edges of roofs, you have provided protection against fires;[14] for they do not catch fire and cannot be carbonized. §2 Oak is durable if buried in underground structures, and to some extent as stakes. *Aesculus* oak is suitable timber for buildings and vine-stakes. Chestnut with its wonderful density lasts in the fields and in roofs and other interior work, only its weight being a fault.[15] Beech is useful in dry conditions, but rots in moisture. Each of the poplars,[16] and willow and lime, are indispensable for carvings. Alder is useless for construction, but indispensable if a moist place needs to have piles sunk in it to receive foundations. Elm and ash become rigid if dried; bendable before, they are considered useful for ties. §3 Hornbeam is very useful, cypress outstanding, pine not durable except in dry conditions. I learnt in Sardinia that steps can be taken to prevent its quick decay as follows; cut beams of it are submerged for a whole year in any kind of pond, to be used thereafter, or else they are buried in sand on the shore, so the advancing waves on the alternating tides can wash against the bank in which they are stowed. Cedar is durable, unless in contact with moisture. Any timber cut from a southern exposure is more useful, while that from a north-facing slope is taller but easily decays.

16. *Transplanting older trees*

This month in dry, hot, sunny places we shall transplant larger trees, with their branches pruned but their roots undamaged; they need to be helped with plenty of manure and water.

---

14. i.e. the spread of fire from neighbouring buildings.
15. Perhaps the first mention of the use of Spanish chestnut in roofs, which was to become so important in medieval buildings (Plommer 1973: 7).
16. Black and white.

Olive Oil

17. *Pressing (Greek recommendations)*

§1 The Greeks in their precepts for making olive oil give the following advice. We should pick only as many olives as we can press that night. For the first oil there should be only light pressure from the millstone. For the pits turn a dirty colour once cracked; so the first pressing should be from the flesh only. The baskets should be made of willow withes, because this species is said to enhance the oil. The oil that flows of its own accord will be finer. §2 Then they advise that salt and *nitrum* should be mixed with the new oil, to redeem its thickness. Next, when the *amurca* has settled, the pure oil should be transferred after 30 days to glass vessels. The second pressing should be made by a similar system, but bruised with more pressure from the mill.

18. *Oil like Liburnian* [17]

First-press oil becomes similar to Liburnian, according to the Greeks, if in the best green oil you mix dried elecampane and laurel leaves and galingale, all ground together and finely sifted, along with baked ground salt, and stir them for some time once added to the oil; then after three days or a little longer, once it has settled, use it.

19. *Cleaning dirty oil*

If oil is dirty, they advise that roasted salt, still hot, should be thrown in and the jar carefully covered; this leaves the oil clean after a short time.

20. *Foul-smelling oil*

§1 If it has a nasty smell, they advise that green olives without the pits be crushed and two *choenicae* be placed in a *metreta* of the oil.[18] If olives are lacking, particularly tender stalks of olive should be crushed similarly. Some mix both of these, adding salt as well. They suspend

---

17. Oil from the Liburnian region on the north-eastern Adriatic coast (currently Croatia) was considered among the best available.

18. A *choenica* is equivalent to two *sextarii* (see footnote on 11.14.5), and a *metreta* is about 71 *sextarii*.

all of this in a linen cloth, and plunge it like that in the vessel of oil. Then after three days have passed they remove it and transfer the oil to other vessels. §2 Some put in an old brick that has been heated. Many plunge in barley loaves, cut in small shapes and wrapped in an open-weave linen cloth, and replace them at intervals with new ones. After doing this two or three times, they toss in salt, transfer it to other vessels and let it settle for a few days.

§3 If some animal happens to fall in and spoil the oil with a putrid stench, the Greeks advise that a handful of coriander should be suspended in a *metreta* of the oil, and remain there a few days. If it does not dispel the nastiness at all, the coriander should be changed, until this defect is overcome. But it will be particularly helpful to transfer it to clean vessels after six days; better still, if they have previously carried vinegar. §4 Some mix in dried ground fenugreek seeds, or repeatedly set olive wood charcoal alight and extinguish it in the oil itself. If there is a bitter odour, they advise that the waste part of grapes, which the Greeks call *gigarta*, should be ground, reduced to a wad, and immersed.

21. *Curing rancid oil*

Rancid oil, the Greeks state, can be cured thus. They advise that white wax, melted in clean best-quality oil, should be placed in it while still liquid; then roasted salt, still hot, should be added, and the whole covered and gypsum-sealed. The oil is cleansed in this way, with both flavour and odour changed. All oil, they say, should be kept in underground places; its nature is such that it is cleansed by sun or fire, or by boiling water mixed together with it in a vessel.

22. *Seasoning olives*

§1 This month too we shall season olives. There are various types. 'Swimming' olives are made as follows. On each layer of olives in turn you sprinkle pennyroyal and pour on moderate quantities of honey and vinegar and salt as an intervening coating. Or again you will spread the olives on top of twigs of fennel or dill or mastic, and after laying little olive branches you pour on a *hemina* of vinegar and some brine, and allow such built-up layers to rise to fill the vessel.

§2 Another method: you will mellow selected olives in brine, and after 40 days pour off all the brine; then you will add two parts of *defritum*, one part of vinegar and finely chopped mint to the vessel and re-cover the olives, so it floats above them and soaks them properly.

Another method: you will leave hand-picked olives for a whole night in the heat of the bath-building, on a board or rack. In the morning you will remove them from the bath, sprinkle them with ground salt, and use. These cannot be kept longer than eight days.

§3 Another method: you first place undamaged olives in brine. After 40 days you will lift them out and slit them with a sharpened cane; if you want them sweeter, you will need to pour on two parts of *sapa* and one of vinegar; if sharper, two parts of vinegar and one of *sapa*.

Another method: take one *sextarius* of raisin wine, well-sieved ash (as much as both hands will hold), one half-*sicilicus* [19] of old wine and some cypress leaves. After mixing all this you pour it on the olives, press them down and steep them by making a coating at intervals, until you reach the tops of the vessels.

§4 Another method: you collect the olives you find lying on the ground puckered with constricting wrinkles, sprinkle them with ground salt and spread them out till they shrivel in the sun. Then you will arrange several successive layers of berries with laurel spread beneath each. Then you will bring *defritum* with a bunch of savory to the boil two or three times; and once it is tepid, after adding a little salt and throwing in a bunch of oregano, you pour the whole mixture over the olives.

§5 Another method: when berries are gathered from the tree you will preserve them straightway. You will spread rue and rock-parsley between the built-up levels, and cover them at intervals with a sprinkling of cumin-flavoured salt. At the end you pour on honey and vinegar. Finally you will mix in whatever amount of best-quality oil.

Another method: you pick black olives from the tree, assemble them and rinse them with brine. Then you place in a pot two parts of honey, one of wine, half of *defritum*. When they have come to the

19. 'One half-*sicilicus*' is perhaps a misunderstanding of αϛ, i.e. 1½ *sextarii*, in Palladius' Greek source: see 8.9 footnote.

boil, you take them off, stir them and mix in vinegar. Once it has cooled, you spread twigs of oregano on top of the olives and pour the whole mixture over them.

§6 Another method: you hand-pick olives with their stalks and sprinkle them with water for three days. Then you place them in brine, remove them after seven days and put them in a vessel with must and vinegar in equal quantities; you will fill the jar and cover it so as to leave some air-passages.

23. *Hours*

Time-reckoning leads November and February through equal hours:

Hour:	1	2	3	4	5	6	7	8	9	10	11
Feet:	27	17	13	10	8	7	8	10	13	17	27

Book 13: December

1. *Sowing*

Now the grains are sown: wheat, emmer and barley (though it is already late for sowing barley). Beans too can be sown around the Septimontium;[1] it is not good to sow them after the winter solstice. Also linseed can still be sown this month, up to December 7th.

2. *Other field tasks*

Now (after mid-month) we shall begin to dig trenched ground for starting vines, as was described earlier (2.10). And this is a good month for us to cut timber. We shall also make posts and baskets and vine-stakes. In cold places we shall make laurel oil, and bruise myrtle and mastic berries in the process of making oil from them, and make myrtle-flavoured wine as mentioned earlier (2.18, 3.27, 3.31).

3. *Gardens*

At this time lettuce should be sown, so it can be transplanted in February. Also now garlic and *ulpicum* and onions and white mustard and summer savory can be sown, by the same regimen and method as was recounted earlier.[2]

---

1. A Roman festival on December 11th.
2. See the index under these various plants.

4. *Fruit trees: hypomelis* [3]

§1 *Hypomelides* are fruits, as Martialis states, resembling service-berry. They grow on a moderate-sized tree with white flowers. The sweetness of this fruit is combined with a sharp flavour. It is sown in December, with the kernels placed in pots. In February the *hypomelis* seedlings, provided they are vigorous and thick as your thumb, are transplanted to a very small hole in loose soil with plenty of dung. §2 But they need to be protected, since they quickly dry out if the wind blows on their roots. They do not reject any kind of soil; they love warm, sunny, maritime places – often rocky ones – but abhor a cold situation. They cannot be grafted, and are short-lived. Their fruit will keep either in jars that have been pitched and covered, or in poplar sawdust, or in grape-jars among the grapes, packed around with marc.[4]

[Paragraph missing][5]

5. *Seasoning turnips*

Now turnips should be cut into small parts and lightly cooked and carefully dehydrated for a whole day, so as to retain no moisture. Then we shall make sure to cover and season them with white mustard mixed in the usual way with vinegar; when the vessels are full we shall close them, and after some days we shall test the turnips for taste and bring them out for use. We can also do this job in January or November.

6. *Preserving food; snaring birds*

Now too those who can harvest the seashore will see to preserving the flesh of sea-urchins in salt, when the waxing of the moon is

---

3. Not certainly identified: one suggestion is *Sorbus torminalis*, the wild service tree.
4. *Vinacea*, the skin, stones, etc. left over from wine-pressing. This was often used as a medium in which to store grapes, as is mentioned by Cato, Varro and Columella.
5. The lost paragraph would have included tasks identified in other months as appropriate for December, viz. planting of chestnuts (12.7.17) and grafting of almonds (2.15.12).

favourable: by its own increase it instructs the bodies of all shellfish and molluscs to swell up. The process is done by the standard method. We can do this job successfully throughout the winter months. We also process hams and bacon, not just this month but in all months that are gripped by winter's cold.

At this time it will be opportune to lay out snares amongst low-growing woods and berry-rich shrubs, to catch thrushes and other birds. This kind of fowling will extend till March.

7. *Hours*

The hours link December with January, but in disparity: though they have similar lines on the sundial, January waxes while December wanes.

Hour:	1	2	3	4	5	6	7	8	9	10	11
Feet:	29	19	15	12	10	9	10	12	15	19	29

Book 14: Veterinary Medicine

1. *The Greeks' advice on care for farm workers*

§1 Against pestilence it is helpful for farm workers who are toiling in the sun's heat to take food briefly and often, so the nourishment provided can refresh them, not accumulate and weigh them down. §2 Most people boil up rue and wild mallow, mix this with a little wine and serve it along with the food. §3 Some think a certain amount of water and milk should be mixed with wine and provided for the workers *before* they eat; they continue this practice from early spring right through to the end of autumn. §4 Some people steep wormwood in hot water and provide this for the workers before they eat – or squill wine or vinegar, but it is better to provide the vinegar *after* food.

§5 Against creatures with spines or poison it will be useful for the farm workers to have access to antitoxin vines; I discussed how to establish these in my farming precepts (3.28). §6 Every shoot of these vines has such healing power that the ash from them easily parries dog-bites – even from rabid dogs in many cases.

2. *Preface*

§1 So there should be no omissions in this work, I have collected together the medical treatments for all kinds of livestock and farm animals, and taken care to lay them out in a single book, with headings designating each and every medical situation,[1] using the

---

1. Manuscript v preserves Palladius' main headings, e.g. *De boum medicina* before chapter 4, and some original sub-headings, including those in chapter 3 and in chapters 40–65: see Rodgers *Introduction* 53–56. Other headings in this translation are mostly my own.

very words of Columella and his sources, so that when need arises the remedies for the pressing situation may be easily be found.

§2 I have also made a short summary of all drugs, so the master can store them all in his house before they are needed, to avoid anything being unavailable when required.[2]

3 §1 *Herbs*

∞ wormwood
∞ caper leaves
∞ horehound
∞ savin
∞ twigs of white bryony
∞ creeping thyme
∞ squill
∞ fennel
∞ the root that shepherds call *consiligo*
∞ mistletoe leaves
∞ papyrus
∞ reed
∞ wild santolina
∞ summer savory
∞ knotgrass or *polygonum*
∞ cow savory
∞ sage
∞ sea lettuce or *titymallus*
∞ burdock
∞ the mountain trefoil with pointed leaves (or a store of its seed)
∞ wild carrot seed
∞ sea wormwood
∞ stavesacre
∞ ivy roots
∞ nettle seeds

2. Within each category (e.g. 'Herbs'), Palladius lists items in the order in which they appear in the following chapters of Book 14. Consequently this summary could serve in part as an index, in addition to being a checklist. I have offered identifications of some of the more obscure herbs and other items in footnotes to the following chapters.

- ∞ plantain
- ∞ mullein
- ∞ henbane (i.e. stinking nightshade)
- ∞ and its seed
- ∞ alfalfa
- ∞ hemlock juice (collected in spring before the plant forms seeds, and kept in earthenware pots)
- ∞ madder
- ∞ giant fennel
- ∞ roots of cane and whitethorn
- ∞ wild beet
- ∞ stinkwort
- ∞ calamint
- ∞ wild cucumber
- ∞ savory
- ∞ mugwort
- ∞ wormwood
- ∞ black nightshade
- ∞ purslane
- ∞ oregano
- ∞ germander

§2 *Fruit tree products* [3]
- ∞ leaves of myrtle
- ∞ cypress
- ∞ mastic
- ∞ wild olive
- ∞ fruits of wild fig
- ∞ laurel
- ∞ pomegranates before they form seeds
- ∞ a pine torch
- ∞ cypress balls

3. The pine torch is a surprising item in this category, but Palladius includes pines among fruit trees (for their kernels) at 12.7.9. Cypress balls are the round cones.

§3 *Pharmaceuticals*

∞ frankincense
∞ cassia
∞ myrrh
∞ gall-nut
∞ pitch
∞ fish sauce
∞ hyssop
∞ sulphur
∞ split alum
∞ litharge
∞ pine bark
∞ Cimolian earth
∞ cumin
∞ rock salt (Spanish or ammoniac or Cappadocian)
∞ silphium (i.e. asafoetida)
∞ bitumen
∞ alum
∞ *nitrum*
∞ galingale

§4 *Articles*

∞ snakeskin
∞ *amurca*
∞ blood of a sea turtle
∞ a duck or marsh birds
∞ aged bovine urine
∞ liquid pitch
∞ goat and ox tallow
∞ aged human urine
∞ old grease
∞ a sponge
∞ raisin wine
∞ butter
∞ honey
∞ aged cheese

- ∞ ox marrow
- ∞ a spider-mouse drowned in olive-oil
- ∞ mastic oil
- ∞ seawater
- ∞ seal fat
- ∞ wax
- ∞ a cuttlefish shell
- ∞ unsmoked honey or thyme honey
- ∞ verdigris
- ∞ cotton
- ∞ *defritum*

§5 *Seeds*

- ∞ bitter vetch
- ∞ marc
- ∞ barley groats
- ∞ barley flour
- ∞ lentil flour
- ∞ linseed
- ∞ millet
- ∞ bitter lupine
- ∞ dry fig
- ∞ git
- ∞ garlic
- ∞ *ulpicum*
- ∞ onions

Treating Cattle[4]

Preventive medicine

4 §1 So cattle can be healthy and maintain their strength, supply them over three days with a generous dose of medication made with

---

4. The Latin *boues* (singular *bos*) covers all domesticated bovines, i.e. both oxen used for ploughing and hauling, and cattle used chiefly for meat. I have generally translated *boues* as 'cattle' and *bos* as 'the animal', but I have used 'oxen/ox' where the reference is clearly to working animals.

ground-up leaves of caper, wild myrtle and cypress in equal weights, mixed with water and stored in the open for one night. This should be done four times a year, at the end of spring, summer, autumn and winter. §2 Often, too, lethargy and nausea are dispelled if you put a whole uncooked hen's egg down their throat when they have not eaten, and the next day grind some garlic or *ulpicum* cloves with wine and pour this down their nose.

§3 These are not the only remedies that lead to good health. Many people mix generous amounts of salt into their fodder; some grind up horehound with oil and wine, others blades of leek; others crush grains of frankincense, others again savin and rue, dilute them with unwatered wine, and give the animals these medications to drink. §4 Many use white bryony stalks and bean husks to medicate cattle. Some grind up a snakeskin and mix with wine. Creeping thyme ground up with sweet wine also serves as a remedy, as does squill cut up and soaked in water. All the above-mentioned potions, given in doses of three *heminae* per day over three days, purge the stomach and restore strength by getting rid of disorders. §5 *Amurca*, however, is held to be particularly healthful, if it is mixed in equal quantities with water and the animals are accustomed to it. It cannot be given straight off: first their food is sprinkled with it, then their water is medicated with a small amount, next it is mixed in equal amounts and given to them to drink at will.

5 §1 At no season, least of all in summer, is it beneficial for cattle to be run, since this causes either diarrhoea or fever. One must also guard against any pig or hen sneaking close to their stalls. For a hen's droppings, if mixed with the fodder, are fatal to cattle, while a sick pig is capable of causing pestilence.

Pestilence and remedies
§2 If pestilence falls on the herd, a change must hastily be made in the climate and area; one must split the animals up into groups and send them to far-off districts, and in this way separate the healthy ones from the sick, to prevent contact with an animal who might infect and impair the others. And when they are trekked out, they must be led to places where there are no livestock pastured, lest their arrival sicken them too.

But all illnesses, however pernicious, must be conquered and driven off with well-chosen remedies. §3 For dropsical animals roots of eryngium[5] should be mixed with fennel seeds, added to flour made from roasted milled wheat, and sprinkled with boiling water, and this medication should be hand-fed like mash to the sick cattle. §4 Then a potion made from equal amounts by weight of cassia, myrrh and frankincense, with the same quantity of sea turtle blood, is mixed with three *sextarii* of old wine, and this is poured through the nostrils. It will suffice to divide the actual medication into equal doses of one and a half ounces, and to give one daily for three days with wine.

§5 We have learnt of an effective remedy from the root that herdsmen call *consiligo*. It grows abundantly in the Marsian mountains, and is extremely good for the health of all farm animals.[6] It is dug up with the left hand before sunrise, for it is believed to be more potent when so gathered. §6 Its use is described as follows. With a bronze pin a circle is drawn on the widest part of the ear-flap, so that as the blood oozes out a shape appears like the letter O. This is done on both the inside and the outside of the ear, and then the centre of the inscribed circle is pierced through with the same pin, and the above-mentioned root is inserted in the hole. As the fresh wound closes on it, it grips it so tightly that it cannot slip out. Then all the virulence and malignancy of the disease is drawn into that ear, until the part that was circumscribed by the pin necrotizes and drops out.

§7 Cornelius Celsus also recommends[7] grinding up mistletoe leaves with wine and infusing this through the nose. This remedy should be used if the cattle are suffering as a herd, and those above if individuals are suffering.

---

5. Columella 6.5.2 wrote *tuncpanaciseterungiiradices*, 'Then roots of all-heal and eryngium ... ' Probably through some corruption in his manuscript of Columella, Palladius interpreted this as *tympaniciseringiradices*, 'For dropsical animals roots of eryngium ...' This eryngium is probably *E. maritimum*.
6. Proposed identification of this plant are *Helleborus viridis* or *Pulmonaria officinalis*, lungwort. The inhabitants of the Marsian region in central Italy were famous for magic.
7. Palladius' references to Celsus all come from Columella. Celsus wrote on agriculture and other subjects, though only his work on human medicine has survived.

Digestive disorders

6 §1 Indigestion is indicated by frequent belches and stomach rumbles, distaste for food, muscle tension and dullness in the eyes; all this causes the animal to stop chewing its cud and cleaning itself with its tongue. The remedy is two *congii* of hot water and then 30 moderate-sized cabbage stalks cooked and given dipped in vinegar; for one day the animal should be kept off other food. §2 Some people keep it shut up indoors, so it cannot graze. Then they mash up four *librae* of top-growth of mastic and wild olive with a *libra* of honey, mix this with a *congius* of water, let it stand outside one night, and pour it down the animal's throat. Next, after an interval, they soak four *librae* of bitter vetch and offer it, giving nothing else to drink. §3 It suits to do this for three days, to dispel all the listlessness. For if indigestion is neglected, the result is that the stomach becomes inflated and there is worse pain in the intestines, which does not allow the animal to take food or stand still, causes groaning, and often forces it to lie down, roll around and wave its tail repeatedly.

§4 An evident remedy is to bind tightly the part of the tail closest to the buttocks, pour a *sextarius* of wine with a *hemina* of oil down the throat, and then drive the animal along at speed for a mile and a half. §5 If the pain remains, one should cut round the hooves, insert a greased hand in the rectum and draw out the dung, then make the animal run again. §6 If this too fails, three dried wild figs are ground up and given with a *dodrans*[8] of hot water. If even this medication is unsuccessful, two *librae* of wild myrtle leaves are pulverized, mixed with two *sextarii* of hot water and poured down the throat through a wooden vessel, and then blood is let under the tail. When enough has flowed, it is staunched with a papyrus bandage. Then the animal is made to run until it is panting.

§7 There are also the following remedies before blood is let. Half a *libra* of crushed garlic is mixed with three *heminae* of wine and after drinking it the animal is made to run, or a *sextans*[9] of salt is ground up with 10 onions, mixed with boiled-down honey and

8. Three-quarters of a *sextarius*, i.e. 410 ml.
9. The manuscripts have a *sextarius*, which is far too much; from Columella 6.6.5 I have restored *sextans*, which is one-sixth of a *libra*, i.e. 55 g or 2 oz.

inserted as a suppository in the rectum, and then the animal is driven at a run.

7 §1 Pain in the stomach and intestines is also calmed by the sight of swimming birds, especially a duck. If an animal with intestinal pain catches sight of a duck, it is quickly relieved of its torment. The sight of a duck also cures mules and horses with even greater success.

§2 But sometimes no treatment avails, and an intestinal infection results. The symptom of this is a bloody mucous discharge from the stomach, and the remedy consists of 15 *librae* of cypress and the same of gall-nuts[10] plus the weight of both in very old cheese. These are ground together and four *sextarii* of dry wine are mixed in. This is given over four days, portioned out in equal amounts, and some fodder is provided, consisting of top-growth of mastic, myrtle and wild olive.

§3 Diarrhoea saps both physique and strength, making animals useless for work. When it occurs, the animal should be prevented from drinking for two days and kept off food for the first day, but then given top-growth of wild olive and reeds, plus mastic and myrtle berries, but only the most sparing access to water. §4 Some people crush tender laurel and wild santolina in two *sextarii* of hot water and pour this down the throat, and offer the same fodder that we talked of above. §5 Some bake two *librae* of marc, then grind it up with two *sextarii* of dry wine, and provide this medication to be drunk, removing all other liquids but nonetheless offering top-growth from the above-mentioned trees.

Aversion to food
§6 If neither a runny discharge from the stomach nor intestinal pain is present, but the animal rejects food, droops its head and often shuts its eyes, while tears flow from the eyes and mucus from the nostrils, the centre of the forehead must be cauterized down to the bone, and the ears slit with a knife. While the wounds made by cautery are healing, it is helpful to rub them with aged bovine urine, but the parts cut by the knife are better healed with pitch and olive oil.

10. A corruption, probably earlier than Palladius, of Columella's '15 cypress cones and the same number of gall-nuts.'

8 §1 Aversion to food is also frequently caused by abnormal growths on the tongue, called 'frogs' by veterinarians. These are cut away with the knife and the wounds rubbed with salt ground up with an equal amount of garlic, until a flow of pus is drawn out. Then the mouth is rinsed with wine, and after an interval of an hour, fresh grass or leaves are given, until such time as the sores scab over.

§2 If there are neither 'frogs' nor runny bowels, and nevertheless the animal is not keen on food, it will be helpful to pour garlic pounded with oil through its nostrils, or rub the throat with salt and summer savory, or smear that area with crushed garlic and fish sauce – provided that the only symptom is aversion to food.

§3 But if there is an attack of bile and nausea (both of which usually trouble cattle in summer), causing an animal to reject food, the right treatment will be to withhold water for one day in hot weather and for two days in cold, then to sprinkle *sextarii*[11] of barley groats with strained fish sauce and hand-feed this as one would to a mule, and gather grass and greenstuff and provide water copiously, so the bowels can be loosened and purged more easily. §4 If even this device does not dispel the nausea and aversion to food, dried human dung is ground up, diluted with water, kept outside for one night and poured down the throat, while all other drink is withheld. Then the animal receives bitter vetch moistened and mixed with chopped-up grass, and subsequently santolina ground up in wine; so it expels the bile, and with it the aversion to food.

Fever

9 §1 A feverish animal should be kept off food for one day. Next day, before it eats, a little blood should be let under the tail. After an hour it should be hand-fed 30 medium-size cooked cabbage stalks, dipped in oil and fish sauce, and these rations should be given for five days on an empty stomach. §2 In addition top-growth of mastic or olive trees should be offered, or exceptionally tender leaves of any kind, plus vine foliage; then too one should wipe the lips with a sponge and provide cold water to drink thrice daily. This treatment should take place indoors, and the animal should not be let out until healthy. §3

11. The number of *sextarii* has been accidentally omitted by the scribes; we cannot restore it since this paragraph is taken from an unknown source.

The symptoms of fever are running tears, a heavy head, tight-shut eyes, saliva flowing from the mouth, slow breathing that is somewhat obstructed, sometimes with groaning also.

Coughing

10 §1 A cough that started recently is best dispelled with a mash made of barley flour. Sometimes grass chopped up and mixed with bean meal is more helpful. Also two *sextarii* of lentils, shelled and milled fine, are mixed with hot water, and the resulting broth is infused through a horn.[12]

§2 A long-standing cough is cured by two *librae* of hyssop soaked in three *sextarii* of water. This medicinal plant is ground up and hand-fed along with four *sextarii* of lentils (milled fine as mentioned) like mash, and afterwards the hyssop water is infused through a horn. §3 Juice of leek mixed with olive oil, or leek leaves pounded with barley flour, serve as a remedy; and leek roots, carefully washed and pounded with wheat meal and given on an empty stomach, dispel a cough of very long standing. §4 The same result is achieved by bitter vetch without the shells, milled with an equal portion of roasted barley and put down the throat like mash.

Abscess

11 It is better to cut open an abscess with the knife than to treat it with medication. Then, when the matter has been squeezed out, the cavity that contained it is washed out with hot bovine urine, and treated with bandages impregnated with liquid pitch and olive oil. If the area cannot be bound up, goat or bovine tallow is dripped into the cavity from a red-hot metal plate. Some people cauterize the affected area, wash it out with aged human urine, and then smear it with liquid pitch and old grease heated up in equal amounts.

Lameness

12 §1 Drainage of blood into the feet causes lameness. When this happens, inspect the hoof straightway. A touch shows the presence of heat, and the animal does not tolerate firm pressure on the affected

---

12. i.e. the narrow end of the horn is inserted in the animal's throat, and the liquid poured in through the broad end.

part. But if the blood is still in the legs, above the hooves, it can be dispersed by persistent rubbing, or, if this is not effective, let out by scarification. §2 If it is already in the hooves, a slight opening is made with a small knife between the two halves of the hoof, and afterwards bandages soaked in salt and vinegar are applied; then the foot is placed in a sandal made of broom, and the greatest care is taken that the animal should be stabled in dry conditions and should not step in water.

§3 If the blood is not let out it will create a swelling, and if this turns into an abscess, it will be slow to heal. First the area is cut round with a knife and cleaned out, then packed with cloths soaked in vinegar, salt and oil, then brought back to health with dressings of old grease and goat tallow cooked up in equal quantities.

§4 If the blood is in the front of the hoof, the point of the hoof itself is cut back to the quick and the blood released here, after which the foot is bandaged and protected with a sandal. It is not useful to open the middle of the hoof at the back, unless an abscess has already formed in that area.

§5 If the lameness is due to muscle pain, the knees, backs of the knees and legs should be massaged until healthy. §6 If the knees are swollen, they should be rubbed with hot vinegar, and one should make a poultice of linseed together with millet that has been ground and sprinkled with honeywater. A further correct treatment: sponges soaked in boiling water, squeezed out and smeared with honey are applied to the knees and wrapped on with a bandage. §7 If there is some fluid under the swelling, yeast or barley flour boiled in raisin wine or honeywater is applied; then, when the abscess has ripened, it is cut open with a knife, drained, and treated with bandages as indicated above. §8 Other things that can heal the area that has been lanced are, as Cornelius Celsus advised, lily root or squill with salt or knotgrass (which the Greeks call *polygonum*) or horehound.

§9 In general, however, all bodily pain (unless caused by a wound) is best dispelled with poultices if it is recent, but with cautery if long-lasting, with butter or liquid goat-fat dripped onto the cauterized area.

Skin diseases

13 §1 The scab is relieved if crushed garlic is rubbed on it. §2 The same remedy cures the bite of a rabid dog or wolf; such a bite, however, is equally well healed if matured pickled fish is laid on the wound. §3 But for scab another treatment is more effective: cow savory[13] and sulphur are ground up together, mixed with *amurca*, olive oil and vinegar, and cooked. When the mixture is hot, ground-up split alum[14] is sprinkled on. This medication is most effective when smeared on in blazing sunshine.

§4 Sores are remedied by ground-up galls, and equally by horehound juice mixed with soot.

Hide-binding

§5 There is also a dangerous affliction for bovine livestock (country folk call it being hidebound): the skin sticks so tightly to the back that when gripped by the hands it cannot be pulled away from the ribs. §6 This condition occurs only if an ox has been reduced to thinness by some ailment, or sweated while working and became chilled, or was rained on while pulling a load. §7 Since these situations are injurious, one must ensure that when oxen return from work still heated and panting, they are sprinkled with wine and cakes of fat are put down their throats. §8 But if the above-mentioned malady has taken hold, it will be helpful to boil down laurel and to use this hot water to rub their backs, then immediately to knead them with plenty of oil and wine, then grip the skin all over and pull it up. This is best done in the open under a blazing sun. §9 Some people mix oil-lees with wine and fat, and use this medication after the above-mentioned dressings.

Ulcerated lungs

14 §1 It is also a serious menace if the lungs become ulcerated; as a result of it, the animal is afflicted with coughing, thinness and finally tuberculosis. §2 To prevent these conditions being fatal, a hole is cut in the ear, as was instructed above (14.5.6), and *consiligo* root is

13. Identified as wild marjoram, *Oreganum vulgare*.
14. 'Split' refers not to processing but to the natural form of a type of alum, which divides into filaments as it develops.

inserted in it; in addition, leek juice, about a *hemina*, is mixed with an equal amount of olive oil, and given as a drink daily for several days along with a *sextarius* of wine.

Neck ailments

§3 If the animal's neck is bruised by work, a very effective remedy is to let blood from the ear, or, if that is not done, to grind up the herb called sage with salt, and place it on the bruise. §4 If the neck is put out and droops, we shall observe the direction in which it is turned, and draw blood from the ear on the opposite side. Next, the largest vein visible in the ear is given a preparatory whipping with a stick, and then, when it swells under the blows, it is lanced; next day blood is let again from the same place, and a two-day respite from work is given. Then on the third day a light task is imposed, and the animal is gradually brought up to its regular duties.

§5 However, if the neck is not bent to either side but is swollen in the middle, blood is let from both ears. If it is not let out within three days of the ox developing the ailment, the neck begins to swell, the muscles tighten, and the consequent stiffness cannot tolerate the yoke. §6 An apt ointment for this kind of ailment, as we have learnt, is made from liquid pitch, bovine marrow, goat tallow, old grease and old olive oil, compounded in equal amounts by weight and brought to a boil. §7 This compound should be used as follows: when the ox is being unyoked from work, the swelling on his neck should be moistened with water and kneaded at the trough from which he is drinking, and then rubbed down and coated with the above-mentioned ointment.

§8 If he completely rejects the yoke because of the swelling, he should be given a few days' rest from work; then the neck should be rubbed with cold water and coated with litharge. Celsus for his part recommends that the herb called sage, mentioned above, should be pounded up and placed on the swollen neck.

§9 The warts that frequently infest the neck are less of a nuisance, for they are easily healed by dripping oil onto them from a burning lamp. §10 The preferable system, however, is to ensure that they do not form, and that the animals' necks do not lose their hair. Their necks only become hairless if wetted with sweat or rain in the course of work. So when this happens, their necks should be dusted with old ground-up brick before they are unyoked, and then oiled immediately after they have dried.

Injuries to hooves and legs

15 §1 If he injures his ankle or hoof on the ploughshare, apply hard pitch and grease together with sulphur, wrap it in grease wool[15] and cauterize above the wound with a red-hot iron. §2 The same remedy is very successful if he happens to tread on a stick (which should be pulled out before treatment), or perforates his hoof on a sharp shard or stone. But if the wound is quite deep, a fairly wide cut is made around it with the knife, and it is then cauterized as advised above. Thereafter a shoe made of broom is worn for three days to protect it, and vinegar is poured on the wound.

§3 Again, if he wounds his leg on the ploughshare, sea lettuce (which the Greeks call *titymallus*)[16] with an admixture of salt is placed on the area.

§4 Feet that are worn underneath are washed with heated bovine urine. Then a bundle of vine prunings is burnt, and after the flames have reduced it to ash, the ox is forced to stand in the hot ash, and his hooves are smeared with liquid pitch plus olive oil or grease.

§5 However, the beasts will be less likely to go lame if, after they are unyoked from work, their feet are washed in plenty of cold water, and then the hocks, the coronets, and the clefts that divide the hooves, are rubbed with old grease.

Other injuries

16 §1 It often happens, too, that an ox wrenches its shoulders, either through the strain of an over-long journey, or in first-ploughing when it has struggled with hard soil or an obstructive root. When this happens, blood should be let from the forelegs: if he has hurt his right shoulder, from the left; if the left, from the right; if he has damaged both quite seriously, veins in the hind legs are opened too.

§2 If he breaks his horns, lint soaked in salt, vinegar and oil is placed on them, they are bandaged, and the same mixture is poured

15. Wool that has not been scoured, containing a high level of lanolin.
16. Sea spurge, *Euphorbia paralias*.

on the bandages for three days. Then on day four grease with equal amounts of liquid pitch and pine bark is reduced to a paste and applied; finally, when they are scarring over, soot is rubbed on them.

Worms

§3 Sores, if neglected, frequently teem with worms. If cold water is poured on them early in the day, they contract with the chill and fall out. Or if they cannot be removed in this way, horehound or leek is ground up and applied with an admixture of salt. This very quickly kills the creatures in question. §4 But once the sores have been cleared out, bandages should be applied straightway with pitch and oil and old grease, and the areas around the wound should be coated with the same ointment, to prevent their being infested with flies, which settle on sores and generate worms.

17 §1 But since there are many animals that cannot be touched because of their fierce temperament or unbroken nature, the following remedy is helpful against worms by which they have been infested. The remedy is brought about as follows against the worms that tend to occur in animals' wounds; it is effective even if the animals are elsewhere, however far distant, and for all kinds of livestock. §2 You will do thus: in the period before sunrise, before you empty your stomach or urinate, you will squat down with feet wide apart, pick up some dust that you find before you or some manure from the stalls in your left hand first, cast it behind you between your feet, and say 'As I cast this, so may the worms be cast from so-and-so's horse' (adding 'white' or 'black' or whatever colour it is). You cast again with your right hand and say the same, and again with your left hand, saying as above. This avails an ox, a horse or any livestock.

§3 Again, another spell has similarly proved effective for the same purpose. Early, before sunrise, you cut a bramble from where it is growing and say, 'As I have cut this, so may the worms be cut from so-and-so's horse or ox' (adding 'parti-coloured' or 'white' or whatever colour it is).[17]

---

17. Chapter 17 is not taken from Columella, but its source is unknown. The spells quoted are full of non-literary Latin, e.g. *caballus* for 'horse', *ille* for 'so-and-so'.

Poisonous bites

18 §1 Snakebite is also fatal to cattle, and the poison secreted by small animals is also harmful. Often, when a beast lies down unwarily in the pasture on a viper or blindworm, they are provoked by his weight and inflict a bite. And the spider-mouse (which the Greeks call *mygale*),[18] although it has small teeth, causes no small trauma. The viper's poison is dispelled if you scarify the area with a knife, pound up the herb called burdock, and apply it with salt. §2 Even more helpful is the crushed root of the same herb, or the mountain trefoil if it can be found. The latter is most efficacious when growing in broken ground. It has a foul smell not unlike bitumen, and for that reason the Greeks name it *asphaltites*, but on account of its shape we call it pointed trefoil:[19] it grows long hairy leaves, and makes a sturdier stalk than meadow trefoil. §3 The juice of this herb, mixed with wine, is poured down the animal's throat, and its leaves, ground up with salt, are used as a poultice. §4 If the green herb is not in season, the collected seeds are pulverized in wine, and given as a drink. Also the roots are ground up with the stalk, mixed with barley flour and salt, moistened with honeywater and applied to the area that has been scarified. §5 Another effective remedy is if you mash up tender top-growth of ash trees, five *librae* weight, with five *sextarii* of wine and two of olive oil, and pour the extracted liquid down the animal's throat, and again pound up top-growth of the same tree with salt and apply it to the wounded area.

§6 The bite of the blindworm causes swelling and formation of an abscess; a spider-mouse bite has the same effect. But the harm caused by the former is cured with a bronze pin, if you puncture the wounded area and smear it with Cimolian earth[20] in vinegar. §7 As for the mouse, it pays with its own body for the malady it has caused, for the creature itself is killed by being drowned in olive-oil, and after putrifying it is ground up, and this medication is used to anoint a spider-mouse bite whenever it occurs. Or, if this is not available and the swelling shows the injury inflicted by the teeth, cumin is

18. Possibly, though not certainly, a shrew.
19. The description suggests *Psoralea bituminosa*.
20. Cimolite, a kind of fuller's earth, used in vinegar solution for inflammation and tumours in humans also; named for Cimolus, an island in the Aegean.

ground up and a little liquid pitch and grease is then added so as to produce the stickiness needed for a poultice. §8 When this is applied it removes the malady. If the swelling turns into an abscess before it can be dispelled, it is best to cut open the gathering with a red-hot blade, cauterize the affected areas and then smear them with liquid pitch in some oil. §9 It is also a practice to cover a live spider-mouse in potter's clay; when dry, this is hung from the neck of the cattle. This renders the livestock immune to the bite of the spider-mouse.

Eye troubles

19 §1 Eye troubles are generally cured with honey. If the eyes are swollen, honeywater is sprinkled on wheat flour and applied. §2 But if there is a white spot in the eye, rock salt (Spanish or ammoniac or even Cappadocian) relieves the trouble if ground up fine and mixed with honey. Equally effective is cuttlefish shell ground up and puffed through a tube onto the eye thrice daily. §3 Also effective is the root that the Greeks call silphium, but ordinary folk name asafoetida in our usage. One-tenth of its weight in ammoniac salt is added to it, and these ingredients are ground up together and poured into the eye by the same method; or else the plant's root is pounded up and smeared on in mastic oil, thus clearing out the ailment.

§4 Running at the eyes is controlled by sprinkling barley groats with honeywater and applying it on the eyebrows and cheeks. Also wild carrot seeds and juice of wild radish blended with honey soothe pain in the eyes. But whenever honey, or any other sweet liquid, is included in remedies, liquid pitch in oil will need to be smeared round the eye, so it does not become infested with flies. For they (and bees too) fly towards the sweetness and scent of honey and other medications.

Leeches

20 §1 Serious illness is often caused by a leech that has been swallowed in water. Attaching itself to the throat, it sucks blood, and as it grows it prevents the passage of food. If it is in such a difficult place that it cannot be pulled off by hand, insert a tube or reed pipe and pour hot oil through it. When touched by this, the creature quickly falls off. §2 The smell of a burnt bedbug can also be introduced through

a tube: when the bug is placed over a fire and gives off smoke, the tube carries the odour that is generated right to the leech, and the odour dislodges it from its grip. But if it is attached to the stomach or intestine, it is killed by pouring hot vinegar through a horn.

Applicability
§3 Though I have prescribed these medications for use with cattle, most of them can certainly be used successfully with all larger livestock.

Intestinal worms in calves
21 §1 Calves, too, are often troubled by intestinal worms, which generally arise from overeating. Hence one should control the amounts, so they digest properly. If they are already afflicted with this malady, half-cooked lupines are ground up, formed into nuggets and put down the throat like mash. §2 One can also grind up sea wormwood[21] with dried fig and bitter vetch, form it into nuggets and hand-feed it like mash. The same result is achieved by mixing one part grease with three parts hyssop. The juice of horehound and leek can also kill creatures of this kind.

Medicating Horse Stock

Basic causes of illness, and treatment
22 §1 Horses generally contract illnesses from fatigue and heat, sometimes also from a chill or when they do not pass urine punctually, or if they drink quickly while sweating after a gallop, or if, after standing for a long time, they are suddenly galvanized into working or running.

§2 For fatigue the remedy is to pour oil, or fat mixed with wine, down the throat. §3 For a chill, compresses are applied; warm oil is poured on the loins, and the head and spine are rubbed with fat that has been heated or with ointment. §4 If the animal does not pass urine, the remedies are somewhat similar: oil mixed with wine is poured over the groin and kidney area; if this is not successful, a thin suppository of boiled honey and salt is introduced into the

21. *Artemisia maritima*, traditionally used like common wormwood (*A. absinthium*) for expelling internal worms.

urinary tract, or else a live fly or grain of frankincense or suppository of bitumen is inserted in the genitals. §5 These same remedies are used if the genitals are inflamed.

Head pain
§6 Indications of pains in the head: tears flowing, ears flaccid, the whole neck plus the head heavy and bowed towards the ground. §7 In this situation the vein beneath the eye is opened, the mouth is rubbed with hot water and food is withheld on the first day. Next day green grass and a drink of warm water is given with no other food, then old hay or soft straw is provided as bedding; at dusk water is given again and a little barley with vetch stalks, using small portions of food in this way to bring the animal up to its regular diet.

§8 If a horse's jaws are painful, they should be fomented with hot vinegar and rubbed with old grease. If this does not cure the problem, they should be cauterized. The same treatments should be applied to them if they are swollen.

Injuries
§9 If he hurts his shoulders or there is pooling of blood, veins should be opened approximately in the middle of each leg; the blood that flows should be mixed with frankincense dust, and the shoulders smeared with it. To avoid letting more than the right amount of blood, the animal's own dung should be applied to the open veins and fastened in place with bandages. §10 Next day too blood should be taken from the same place and stopped in the same way; the animal should be kept off barley, and given a little hay. Then, after three days and up to day 6, leek juice to about three *cyathi* should be mixed with a *hemina* of oil and poured down the throat through a horn. §11 After day 6 he should be made to walk, but slowly; after walking it will be beneficial to drive him into a pool to have a swim. Thus, as more solid food is gradually introduced, he will be brought up to the regular diet.

Intestinal problems
§12 If bile is troublesome to a horse, so his belly swells and he cannot pass wind, the hand is greased and inserted in the anus to open the

blocked natural exit. Once the dung has been drawn out, cow savory and stavesacre are ground up with salt and mixed with boiled honey, and these ingredients are formed into a suppository which is inserted; as this loosens the bowels, all the bile comes away. §13 Some people grind up a *quadrans*[22] of myrrh and pour it down the throat with a *hemina* of wine, and smear the anus with liquid pitch. Others wash it with seawater, others with freshly made brine.

§14 In addition, maggots and intestinal worms often cause troubles in the gut. Symptoms of their presence are that horses roll around in frequent bouts of pain, or move their heads close to their bellies, or toss their tails quite often. An effective treatment is as described above, i.e. to insert the hand and remove the dung, then clean the anus with seawater.[23]

Cough

23 §1 A recently started cough is cured by pounding lentils, separating them from their pods, and grinding them small in the mill. When this is done, a *sextarius* of hot water is mixed with the same quantity of the lentils, and poured down the throat. The same medication is given for three days, and the sick animal is reinvigorated with green grass and the top-growth of trees. §2 An established cough, on the other hand, is dispelled by pouring down the throat leek juice, three *cyathi*, with a *hemina* of oil daily for several days, and providing the same foods as recommended above.

Skin diseases

24 §1 Patches of impetigo and any rough areas are rubbed with vinegar and alum. Sometimes, if they persist, they are anointed with equal quantities by weight of *nitrum* and split alum mixed with vinegar, and the pustules are scraped with a strigil in the full blaze of the sun to the point where blood is drawn. Then equal quantities of wild ivy root, sulphur and liquid pitch are mixed, together with alum, and the disorders in question are treated with this medication.

---

22. A quarter of a *libra* (Roman pound), i.e. about 82 g or 3 oz.
23. The essential second part of Columella's treatment for worms, viz. giving a vermicidal medicine by mouth, has been omitted here, probably by accident, whether before or after Palladius.

§2 Chafed areas are bathed twice daily in hot water, then rubbed with baked ground salt mixed with fat, till the blood flows.

§3 Scab is fatal to this species, unless help is given quickly. If it is light at the outset, it is anointed repeatedly in bright sunlight with mastic oil, or with nettle seed ground up in olive oil. The most salutary application, however, for this impairment is seal fat. §4 But if it has already become chronic, one needs more powerful remedies. Accordingly bitumen, sulphur, wax, liquid pitch and old grease are mixed in equal quantities by weight and boiled up, and this medication is used for treatment – with the proviso that the scab must first be cut away with the knife and the area rinsed thoroughly with urine. §5 In addition, it has often proved beneficial to cut away the scab to the quick with a scalpel and completely remove it, and to heal the resulting sores with liquid pitch and oil, which clean and refill the wounds. §6 When they have filled, it will be especially helpful, so the sores will scar over and grow hair more quickly, to rub them with soot mixed with new olive oil.

25 §1 As for the flies that infest the wounds, we shall get rid of them by pouring on pitch and oil or grease.

§2 All other sores are properly treated with flour of bitter vetch.

Eyes and nose

§3 Scar-tissue on the eyes is reduced by rubbing with saliva (taken when fasting) and salt, or with cuttlefish shell ground up with rock salt, or with wild carrot seed that has been pounded up and squeezed through a linen cloth over the eye. §4 In addition, any pain in the eyes is quickly remedied by anointing with plantain juice mixed with honey gathered without smoke; if that is not available, thyme honey is second best.

§5 Occasionally even a nose-bleed proves dangerous: it is stopped by pouring the juice from fresh coriander into the nostrils.

Loss of appetite

26 §1 Sometimes too an animal weakens through aversion to food. The remedy for this is the type of grain called git: two *cyathi* of this, ground up, should be thinned in three *cyathi* of oil and a *sextarius* of

wine, and the mixture poured down the throat. §2 As for nausea, it is dispelled if you frequently provide a drink consisting of a head of garlic crushed with a *hemina* of wine.

Abscess

§3 It is better to open an abscess with a red-hot blade rather than a cold knife. After being drained it should be treated with dressings.

Sudden wasting

§4 There is also the well-known and dangerous type of breakdown, whereby in the space of a few days horses are afflicted by sudden wasting followed by death. When an attack occurs, it is helpful according to Columella to pour four *sextarii* of fish sauce through the nostrils of each animal if it is fairly small, but a *congius* if larger.[24] This treatment draws out all the mucus through the nostrils and cleanses the animal.

Narcissism

27 §1 As Columella says[25] (but I have not been able to verify it), mares sometimes spy their reflection in water and fall in love with it. As a result they forget about grazing and die of love and starvation, since their passion for an empty image is equalled by the fire of genuine ardour. §2 The symptom is that they roam headlong through the pastures as if goaded, and repeatedly look around as if searching for something with longing and affection. §3 Cut the animal's mane unevenly and lead her to water, so you can use the very object that caused the affliction to cure it, and avert the dangers of beauty by the benefit of ugliness.[26]

24. Columella 6.34.2.
25. Columella 6.35.
26. According to a surviving passage of Sophocles' lost *Tyro*, when a filly is shorn and sees her reflection in a stream, she is ashamed and grief-stricken. Other authors state that mares become less eager for sex when their manes are shorn (Aristotle *History of Animals* 572b 7–9, Pliny 8.164, Aelian *History of Animals* 11.18).

Treating Mule Stock [27]

28 §1 If a mule is on heat, she is given uncooked cabbage.

§2 If the animal is asthmatic, blood is drawn, and a half-ounce of frankincense and about a *hemina* of horehound juice, mixed with a *sextarius* of wine and of oil, is poured down its throat.

§3 If it is spavined [28] barley flour is applied, and then the abscess is opened with the knife and treated with dressings.

§4 Swellings around the ankles are treated by cutting them open, and sometimes by cauterizing. §5 Blood that has drained into the feet is released as with horses. [29] Or, if there is a supply of the herb that countryfolk call mullein, it is given as fodder. There is also henbane; if its seed is ground up and given to the animal with wine, it cures the ailment in question.

§6 Thinness and lassitude are relieved by frequent administration of a drink: the recipe is half an ounce of sulphur, a raw egg and a denarius weight of myrrh. [30] These ingredients are ground up and mixed with wine, to be poured down the throat. §7 (A cough or stomach-ache is cured equally well by the same ingredients.) For thinness, however, nothing is as effective as alfalfa. This plant when green conditions draft-animals quite quickly, and not too slowly when given dry in place of hay. But it must be given in moderation, to avoid distending the animal with an excess of blood.

§8 If a mule is tired and over-heated, fat is put down its throat, and wine poured into its mouth.

Other treatments for mules will be carried out as we described in the earlier sections of this Book, concerning the care of cattle and horses.

27. Columella begins by saying that he has already described, under other animals, most treatments required by mules, but that he will discuss here certain conditions peculiar to mules (6.38.1). Palladius characteristically omits this general sentence, which is partially repeated in the last sentence of the chapter.
28. i.e. suffering from inflammation of the hock.
29. See 14.12, which however discusses this problem in relation to cattle not horses. The error is Columella's, repeated by Palladius.
30. A denarius is about 4 g or ⅐ oz.

Care of Sheep

Pestilence

29 §1 If the whole flock is sick, we must quit the pastures and watering-places of that entire region, and seek a different climate. If it was heat and sultry weather that caused illness to attack the flock, we must be sure to choose shaded country, but if cold was the cause, we must choose sunny places. §2 Mind, we must follow the flock at a moderate pace, without haste, to avoid aggravating their weakness by long journeys. On the other hand, the droving must not be utterly slow and lazy; §3 for while it is unsuitable to drove sick animals vigorously and strain them, it is conducive to exercise them moderately and rouse them from their lethargy, so to speak – not allow them to sink into torpor and be smothered by it. §4 Then, when the flock has been led to its destination, it should be split up into bands among the tenant farmers, for sheep regain health more easily when separate than en masse, whether because the contamination and exhalation from the disease itself is less in a small number, or because more care can be given more efficiently by reason of numbers.

Scab

30 §1 Sheep are attacked by scab more frequently than any other animal. It generally starts as a result of chilly rain or ice, or after shearing if you do not apply the aforementioned preventive medication (6.8), or if you do not wash off summertime sweat in the sea or a river, or if you allow a newly shorn flock to be wounded by thorn-thickets or brambles, or if you use stalls where mules or horses or donkeys have stood. Above all, however, short rations cause loss of condition, which causes scab. §2 When it starts to sneak in, it is recognized as follows: the animals gnaw at the affected part, or pound it with horn or hoof, or scratch it against a tree, or rub the walls. §3 When you see an animal doing this, you must catch her and part the wool, for the skin underneath is rough and somewhat irritated. Help must be given at the first moment, to avoid contaminating all the stock, since the infection quickly attacks other livestock too, but sheep most of all. §4 There are several treatments, which we shall

list – not that it is necessary to use them all, but in the hope that one of the several may be available and serve as a remedy, since we cannot find some of the ingredients in certain regions.

§5 The mixture that we described just above (6.8) has a satisfactory effect, i.e. the liquid from boiled-down lupines added to wine-dregs and *amurca* in equal amounts. The sap of green hemlock can also remove scab. After being cut down in spring, when it forms a stalk but not yet seeds, it is pounded up, and the juice pressed from it is stored in earthenware vases, with half a *modius* of baked salt mixed into two *urnae*[31] of the liquid. §6 Once this is done, the vase is sealed and buried in the manure-heap. After maturing for a whole year in the heat of the dung, the medication is then brought out, warmed, and rubbed on the scabbed area; this area, however, is first scraped with a rough sherd or rubbed back to the sore with pumice. §7 Another remedy is *amurca* boiled down by two-thirds, or again aged human urine scalded with red-hot tiles. §8 Some people, however, reduce the urine by one-fifth over the fire, mix in an equal quantity of sap from green hemlock, then pour a *sextarius* of crushed salt into each *urna* of this liquid. Again, an equal quantity of ground sulphur and liquid pitch, cooked over a slow fire, does the job. §9 But *The Georgics* declares that no medicine is more effective

'than if a person can cut the ulcer's surface
with the knife; the malady lives and thrives by hiding.'
So one should cut it open, and treat it with medications like other wounds.

§10 Vergil then adds, equally judiciously, that if sheep are feverish, blood should be let from the pastern or between the two parts of the hoof; in the highest degree, he says,

'it helps to dispel the burning heat, and strike
the throbbing vein between the pads of the foot.'[32]
But we shall also let blood under the eyes and from the ears.

Lameness
31 §1 Lameness attacks a sheep in two ways. Either there is chafing and

31. An *urna* (here and in §8) is 24 *sextarii*, i.e. roughly 13 litres.
32. Of Columella's many literary references and quotations, these are the only two to survive into Palladius: they are respectively *Georgics* 3.453–54 and 3.459–60.

morbid discharge right in the division of the hoof, or else the same area contains a boil, near the middle of which a hair projects like a dog's: beneath this there is a worm. §2 The chafing and discharge are quickly cured with pitch that is naturally liquid, or with a mixture of alum, sulphur and vinegar, or with an unripe pomegranate (before it forms seeds) pounded up with alum and steeped in vinegar, or by sprinkling verdigris on the area, or by burning a gall to ash, pulverizing it in dry wine, and smearing this on.

§3 As for the boil containing a worm, we must wield the knife with the greatest caution as we cut around it, lest in the process of severing the boil we should wound the creature inside. §4 For when it is injured it releases a poisonous fluid that soaks the wound, and makes it so incurable that the whole foot must be amputated. After you have carefully cut around the boil, drip hot tallow on the wound by means of a burning torch.

Lung disease

32 §1 The proper cure for a sheep with lung disease is similar to that for a pig: one inserts in the ear the root that veterinarians call *consiligo*. We spoke of it in relating the treatment of larger livestock (14.5.5–6, 14.14.1). §2 This disease is generally contracted in summer if there has been a lack of water; consequently in hot weather all quadrupeds should be given the opportunity to drink large amounts. §3 Celsus' policy if the malady is in the lungs is to give as much sharp vinegar as the sheep can take, or to heat about three *heminae* of aged human urine and pour it through the left nostril with a horn and to put a *sextans*[33] of grease down the throat.

Erysipelas

§4 An incurable condition is the sacred disease, called 'pustule' by shepherds. Unless it is controlled in the first animal afflicted with this malady, it infects and prostrates the entire flock: it does not allow of any remedy, by medication or the knife, for it flares up under almost any touch. §5 The only thing it tolerates is poultices of goat's milk; this infusion somewhat soothes the fiery intensity on contact, but delays rather than prevents the death of the flock. §6 The notable

33. One-sixth of a *libra*, i.e. 55 g or 2 oz.

Egyptian author, a citizen of Mendes,[34] whose handbook is wrongly published under the name of Democritus, recommends that because of this disease one should examine the sheep's backs frequently and carefully: if this malady happens to be discovered on one of them, we must immediately dig a hole at the threshold of the stalls, set the pustulous sheep on its back and bury it alive, then allow the whole flock to pass over the buried animal, because this action drives off the disease.

Miscellaneous conditions

§7 *Bile*, which is no small danger in summer, is dispelled by a draught of aged human urine; this is also a remedy for an animal that has jaundice.[35]

§8 If *catarrh* is troublesome, twigs of summer savory or madder wrapped in cotton are inserted in the nostrils and turned until the sheep sneezes.

§9 *Broken legs* in livestock are treated no differently than in humans, by wrapping them in wool steeped in oil and wine, then binding them up, with a giant fennel stalk as a splint.

§10 Another serious danger is *the herb knotgrass*: if a sheep grazes it, the whole belly is distended, and the animal hunches up, foams at the mouth and excretes thin foul-smelling matter. Blood must quickly be let under the tail, in the area next to the buttocks, and similarly a vein must be opened in the upper lip.

§11 Sheep suffering from *asthma* must be cut in the ear and shifted to a different region – something we recommend in all pestilential diseases.

Lambs

§12 Lambs too must be helped if they are feverish, or weakened by some other illness. When suffering from disease they should not be put in with their mothers, lest they transfer the blight to them. The

34. Columella gave his name, Bolus, but this became corrupted in the MSS to *dolus*: faced with this nonsensical reading, Palladius just calls him *ciuis*, a citizen. Bolus lived in the third century BC: his work *Sympathies and Antipathies* somehow became attributed to Democritus of Abdera.
35. Palladius' copyists make an amusing error here, writing *arguto* 'sharp-voiced, garrulous' in place of the rare *arquato*, 'jaundiced'.

ewes, then, should be milked out separately; their milk should be mixed with an equal amount of rainwater, and given to the feverish lambs to drink. Many people medicate lambs with goat's milk, which is poured down the throat through a horn.

§13 Mouth disease, which shepherds call *ostigo*, is also potentially fatal. It generally occurs if, as a result of the shepherd's carelessness, lambs (or kids too) have got out and have grazed on dew-covered grass, which is interdicted. When this happens, something like erysipelas encircles their mouth and lips with foul sores. The remedy is hyssop and salt ground up in equal amounts by weight: this mixture is used to rub the palate and tongue and the whole mouth. Next the sores are washed with vinegar, then smeared with liquid pitch. Some people's policy is to mix one part of verdigris with two parts of old grease, and to use this heated as a medication.

§14 The system of castration has already been recounted (6.7.1–4), since the same system is followed in lambs as in the larger quadrupeds.

§15 Since enough has been said about sheep, I shall now turn to goat stock.

Treating Goats

33 §1 When other species of livestock are afflicted by pestilence, they first grow weak with sickness and lassitude. Only goats, though fine and lively, suddenly fail, and are laid low en masse as if by some kind of collapse. For this reason, as soon as just a few have been hit with a plague of this kind, they should all be bled; also they should not be grazed all day, but should be shut up in their fold for the four middle hours of the day.

§2 If some other kind of ailment attacks them, they should be medicated with a draft made from the roots of reed and whitethorn. We pound these up thoroughly with an iron pestle, mix in rainwater, and give the animals nothing else to drink. If this does not dispel the sickness, the animals should be sold, or, if even this is not feasible, they should be put to the knife and their flesh preserved with salt.

Then, after leaving an interval, it will be appropriate to replace the flock, but not until the pestilential season has passed – if it was winter, into summer, or if autumn, into spring.

§3 When individual animals are suffering from disease, we shall apply the same remedies as for sheep. When there is subcutaneous water (the ailment that the Greeks call *hydrops*), a light incision should be made in the skin under the flank to drain the morbid fluid, and then the wound should be treated with liquid pitch. §4 When a she-goat's genital area swells after birthing, or the afterbirth does not come away, a *sextarius* of *defritum*, or failing that the same quantity of good wine, should be poured down the throat, and the vulva should be filled with liquid wax ointment. But (to avoid going through the particulars here) we shall medicate goats as has been prescribed above for sheep.

Other Authorities' Advice on Sheep [36]

34 §1 *Against scab* in ovine stock, the sheep should be rubbed with unguent after shearing. But if the scab does take hold, you treat them as follows. Take *amurca* and boiled-down bitter lupine water and white wine dregs: you mix them in equal quantities, heat them in a vessel and rub on the sheep for two days. §2 On day 3 you wash them with seawater or hot brine, and afterwards with fresh water. §3 Others, however, boil up cypress balls [37] with water and rub the sheep with this. Others mix sulphur and galingale with white lead and butter, and rub this on; others rub the sheep with the urine of a donkey that happens to be standing in the road. §4 Others do none of these things for sheep with scab, but shear the affected places and wash them with aged urine. §5 In Arabia they use cedar-resin, as in treating camels.

§6 *Against lice or ticks*, rub the sheep with cedar-resin.

35 §1 *Against vertigo* in sheep: if as a result of the sun's heat they

36. Chapters 34–35 are based on the Greek authorities in the *Geoponika*. The fact that Palladius placed this material here suggests that he regarded Chapter 33 as an appendix to the treatment of sheep.
37. i.e. the round cones.

250

are dizzy, fall down frequently and have no appetite, boil the greens of wild beets and pour the liquid down their throats, and offer the beets in their food.

§2 *Against asthma* in sheep: cut their ears with the knife, and transfer them to another location. These steps are also helpful for goats.

§3 Sheep will maintain their health if, at the beginning of spring, mountain sage and horehound are pounded together and mixed in their drink for 14 days. §4 This should also be done in the same fashion in autumn. If they are sick, tree-medick fodder or cane roots will be also be beneficial.

§5 It is said that a flock of sheep is protected *from wolves*, if squill is hung on the neck of the lead animal.

§6 *Against snakes,* place in their stalls women's hair or galbanum or stag's horn or goat's hooves or bitumen or castor or stinkwort or calamint or santolina or anything with an acrid odour; whether burnt or strewn, they will be effective.

Treating Pigs

36 §1 Symptoms of fever are that the pigs carry their heads sideways at an angle, and that in the pasture, after running forward a little way, they suddenly stop and fall down through dizziness. §2 We should observe which side their heads incline to, so we may let blood from the ear on the opposite side. Again, we must strike the vein on the underside of the tail two fingers from the buttocks – a vein which is quite sizeable at that spot. First it should be beaten with a twig; then, when it swells under the blows from the stick, cut open with the knife; after bloodletting, bound up with willow or elm bark. §3 After doing this we shall keep the animals indoors for one or two days, and provide moderately hot water (as much as they want) plus a *sextarius* each of barley flour.

§4 If their glands are swollen, blood should be let under the tongue. After it has flowed, the right treatment is to rub the whole mouth with salt ground up with wheat flour. Some people think it a more effective remedy to pour three *cyathi* of fish sauce through a horn down the throat of each animal, then split sections of giant fennel, tie linen cords to them, and hang them on the animals' necks, so the fennel is in contact with the glandular swellings.

§5 If they are nauseous, it is thought salutary to mix ivory sawdust with rubbed salt and fine-ground beans, and put this before them when they are hungry before going out to pasture.

§6 It also happens that the entire herd sickens to the point of emaciation; they have no appetite, and when led to pasture they lie down in the middle of the field and snooze in the summer sun, weighed down by a kind of torpor. §7 If this happens, the whole herd is held in a roofed fold, and kept off food and drink for one day. Next day they are given ground-up root of snake cucumber[38] mixed in water, with nothing else to drink. When the animals drink this, they become nauseous, vomit, and are purged. Once all the bile has been expelled, they are allowed to have small chickpea or beans sprinkled with strong brine, and then a drink of hot water as with humans.

37 §1 Pigs should be pastured around streams or lakes and muddy areas in summer, or else ponds should be dug to keep mud available, since if they do not cool their plump swollen bellies frequently with mud, they suffer from the seasonal heat and contract lung disease. This disease in pigs is treated by inserting the little root, i.e. *consiligo*, in the ears.

§2 Often, after consuming immoderate quantities of fruit when seasonal abundance makes it available, they develop a problem with the spleen. In this situation you will fashion troughs from tamarisk trunks, and water the pigs there, for the sap from the wood cures this problem.

38 §1 Pigs are castrated in spring or autumn. The method is to make two incisions and push out the two testicles simultaneously, one through each cut; then the openings are treated with the remedies discussed above on castration (6.7).

§2 We should caution that when sows eat their own young, this is not considered an uncanny event: being quite unable to endure hunger, pigs consume whatever they find, and this is not observed as a portent.

---

38. *Ecballium elaterium*, squirting cucumber, used as a purgative in humans also; usually called 'wild cucumber' in Palladius.

§3 To prevent pigs falling sick, you will give them river crabs to eat.

Advice from Other Greek Authorities

Cattle

Preventive medicine

39 §1 You will make your cattle fat if you give them cabbage stalks soaked in sharp vinegar when they return from pasture, and subsequently sifted chaff mixed with wheat bran for a period of five days; on day 6 you give them four *cotulae* of milled barley, increasing the feed for six days thereafter. §2 In winter they should be fed after cock-crow and again at dawn. §3 Through the winter they should be washed with hot water, but in summer with lukewarm water, and their mouths should be washed with aged urine, to dispel the accumulated phlegm and to purge the tongue of worms: you will also rub the tongue with salt. §4 To protect their health, part of a stag's horn should be hung on them; and to counter any hint of plague the tips of their horns should be bored through and ass's fat inserted. §5 Pigs and hens should be barred from their stalls, since the droppings and feathers are harmful. §6 To prevent their swallowing fruit stones, you will hang up a wolf's tail over the stall. Cattle will not sicken if you boil up galingale roots in wine, or cypress roots, and provide this as a drink.

40 *For headache*, which you recognize if he droops his ears and does not feed, you will rub his tongue with savory ground up in wine and garlic and very fine salt and uncooked pearl barley dissolved in wine. Alternatively you place a handful of laurel leaves in his mouth, and you will need to dissolve a bean-sized piece of myrrh similarly in wine, and pour two *cotulae* of this liquid through the nostrils.

41 §1 *Against colic in cattle*: the symptom is that the animal neither stands still nor grazes. §2 He should be given a moderate amount of food, and we should puncture the flesh near the hooves, to let blood flow. §3 Some people make the puncture around the tail so the blood

can ooze out, and bind on a hyena bone with a piece of cloth. §4 Others mix onions and salt together, push this well within the anal canal, and cause him to run.

42 §1 *Against fever in cattle*: you will take grass, wash it, and offer it as fodder along with vine leaves. §2 The animal should drink very cold water, not in the open but in a shaded place. Its ears and nostrils should be moistened with a sponge soaked in water. §3 In addition, some people cauterize above the forehead and under the eyes, and wipe with a sponge dipped in heated aged urine twice per day, until the scabs from the burning fall off and the wound is healed. §4 The ears too are lanced, to let blood flow. Some people provide barley groats with wine for consumption; some offer tree-medick with wine.

43 §1 *Against coughing in cattle*: you will take soft chaff that has been cleaned and three *cotulae* of ground bitter vetch, and give this in three portions for the animal to eat. §3 Some people grind up mugwort with water, and provide this for seven days, setting it out in the open.

44 *On a suppurated wound*: if pus has gathered, the wound should be cleaned of it, washed with heated aged bovine urine, and wiped with wool. Then a plaster of salt and liquid pitch should be applied.

45 *Against scab and pustules in cattle*: you will rub them with aged bovine urine and butter. Some people mix liquid resin with sulphur, and treat with this.

46 *For an animal with a poor appetite*: sprinkle its fodder with a suitable amount of *amurca*, or smear its horn right down to the roots with oil and resin mixed in equal quantities.

47 *A solution against gad-fly*: pound laurel berries and boil them in water. If you sprinkle this water on the pasture, the flies will be driven away from the land. Alternatively dissolve white lead in water and rub the animals with this, or with ground wormwood.

48 §1 *If bulls are uninterested in mating*: a stag's tail should be burnt,

reduced to powder and mixed with wine, and the bull's genitals and testes should be touched with this ointment. §2 It will also help to stimulate sexual passion in other species. §3 Knotgrass also fecundates all kinds of animals. §4 If cows are uninterested, their genitalia should be touched with squill crushed in water, after they have been cleansed.[39]

Horses

49 §1 *If a horse loses condition*, you will offer him a double ration of roasted wheat and barley, and water him thrice daily. §2 If he remains in the same state, wheat bran should be mixed[40] and gentle exercise employed. §3 If he has no appetite, you soak leaves of black nightshade and germander in water and mix them with the feed; you also infuse bitter vetch in water and make it available, or press two *cyathi* of git, put in three *cyathi* of olive oil, and pour this over the feed with ten ounces of wine.

§4 *Animals with nausea* you treat as follows: you mix garlic and 10 ounces of wine, and give this as a draught.

§5 *If he has difficulty urinating*, you mix the white of 10 eggs with the above-mentioned ingredients, and give this by mouth.

50 §1 *Feverish animals* you will treat as follows. In winter the animal should be warmed in a hot bath, so as not to become chilled. He should receive a moderate feed of bitter vetch or bran plus warm water; his whole body should be rubbed with warmed wine and oil, and his stomach treated.[41] Blood should be drawn from the neck or chest or throat or feet; you will also wash his knees with hot vinegar and, when he is strong, with hot water. §2 If he is fevered and thin as a result of work, you should give 10 ounces of goat milk, a modicum of fine meal, five ounces of olive oil, four eggs and the juice of ground-up purslane daily for three days or more, until he is well. §3 But if the fever arose from rheum in the throat or head, you mix oregano and pitch with oil and rub this all over his head,

39. i.e. after menstruation.
40. Into his feed.
41. i.e. purged.

massage his feet or knees with hot water, rub his mouth with black nightshade ground up with wine lees, and offer him green fodder, if you have it, mixed with barley.

51 *For swelling of the eyes*: if an eye is swollen, mix frankincense, fine meal, lamb marrow (three *scripuli*),[42] double that quantity of rose-flavoured oil, white of two eggs, and smear this on the eye.

52 §1 *Other treatments for swollen eyes*: you smear on frankincense and fine meal mixed with Attic honey, or you add an equal quantity of butter. Or you grind up cuttlefish shell, reduce it to a powder, and puff it onto the eye through a reed pipe; or you smear on root of silphium (i.e. asafoetida) with oil and pitch twice daily.

§2 *For muscle pain*:[43] you bathe his muscles and head with hot water. Then you mix bovine fat, myrrh and sulphur, set them over coals in a cooking-pot, cover his head and bend it over so as to fumigate it with the smoke. Mind, you also treat the stomach and let blood from the tail.

53 *For diarrhoea*: you draw blood from the veins of the head, and give warm water to drink with fine barley flour. Pomegranate rind, ground up and given by mouth, also binds the stomach.

54 §1 *Against colic in a horse*: you wash him with hot water and cover him with a cloth. Presently you put five drachmas of myrrh,[44] six *cotulae* of old wine and three *cotulae* of olive oil through a strainer together, and give this over three days in equal quantities. You also treat his stomach with seawater, with myrtle boiled together in the hot water. §2 If the condition persists, you mix germander leaves and santolina or bitter almonds with dry black wine, or give pomegranate rind with water.

---

42. A *scripulus* is $\frac{1}{24}$ of an ounce, i.e. 1.14 g.
43. I have supplied this heading, missing in the MSS of Palladius, from the Greek sources. The other headings italicized in chapters 40–65 are present in v, and some in M also.
44. A drachma is probably about 4.3 g (see footnote on 8.8.2).

55 *A sprain* you treat as follows.[45] A complaint in the lungs is treated with a draught of very sharp vinegar, heated up and then administered, or human urine.

56 *A starting cough* you treat as follows: you mix fine flour from barley and bitter vetch or beans and give it as a drink. But if the cough has taken hold, you will give two ounces of honey or liquid pitch by mouth.

57 *Treatment of a swelling*: all kinds of swellings will subside if you treat them with salt and olive oil and germander foliage which has been scorched and mixed with wine, all this used as a plaster.

58 *If the horse has swallowed a leech*: lay him on his back and pour hot oil down his throat through a horn.

59 *Against a scorpion sting or the like*: you use pig dung as a plaster on the area that was struck. Take pounded black nightshade, spurge or stinking nightshade or linseed or alum or baked sodium carbonate or rock salt: you will have success by making a plaster from these.

60 *Against difficulty in urinating*: you place the skin of a peeled onion around the genitals. Others grind up parsley seed with wine (two *cotulae*), or give an onion with wine, or pigeon dung, or germander leaves, or a modicum of *nitrum* ground up with a garlic head, or pig dung, mixed with black wine and given through a horn.

61 *If there is blood in the urine*: you mix some cleaned cooked beans with deer fat in a modicum of wine and give it over three days; or you mix goat milk (10 ounces), fine meal, three eggs, and three *cyathi* of olive oil, and administer it though a horn.

62 *To prevent mares from resisting a stallion*, you will grind up *nitrum* and the droppings of smaller birds, together with terebinth resin, and rub their genitals with this or with squill. Once she has been

45. This passage reflects the late antique belief that an *internal* 'sprain' (*uulsum*), caused for example by jumping, could affect the lungs (see Adams 305–08).

mounted, you will keep her away from him for 20 days, and then try her. If she rejects him, you know she is pregnant.

63 *If foals have excessively soft hooves*, you will mix aged fat of pig and he-goat and natural sulphur and garlic, and rub this on the hooves and 'twins'.[46]

64 §1 Their mothers should be kept away from coition, since milk with colostrum is harmful, as with humans.

§2 Again, the hooves are hardened thus: a sherd is heated, vinegar is poured on it, and it is placed under the hooves; and when the animals return after a journey the hooves are washed in cold water.

§3 Horses' bodies benefit from swimming in the sea and rivers. §4 Their coat improves if frequently rubbed with oil and wine.[47]

65 *For diarrhoea* in any animal or in a horse: you write on a papyrus sheet the name *honore panassi*,[48] and fasten it on the top of the tail next to the excretory ring.

46. The horny substances on the soles of the feet, called 'frogs' in English (and sometimes in Latin, *ranulae*) and 'swallows' in Greek.
47. This medley of topics comes from a chapter in the *Hippiatrika* on raising foals.
48. The manuscripts disagree here, and have no doubt mangled the original, which could have been a name or word (*nomen* can mean either), probably magical gobbledegook.

POEM ON GRAFTING

Preface to Pasiphilus

You have here another testimony to the reliance you have bestowed on me: as interest to cover the time elapsed, I have appended for you this work on the art of grafting. But the fact that these volumes on farming were transcribed later than you directed results from the dawdling hand of the copyist. I never make an issue of his slowness, for I know what form the craftiness of servants often takes. I prefer to wait for the results of his work, rather than fear them. Perhaps all masters have this in common, but it has proved difficult for me to find the right balance in slaves' characters. Very often their servile nature spoils whatever worth they may have, and mingles desirable qualities with the opposite. Speediness runs into misbehaviour; slowness takes on a cloak of amiability, and while it eschews wrongdoing, it eschews quickness to the same degree. However, I long deferred speaking of my embarrassment with you; but he has done the job in the style of a good servant.[1]

However, I do not know if your mind is at all inclined to such minutiae as these. It will be some great subject, worthy of ambition, that your studious disposition pursues, and though you may regard these trifles of mine with favour, I do not hesitate to make my own assessment of my resources. People of high standing do not cast their gaze through the dust looking at pennies, since trifling profits are seen as somehow dishonourable in the greatest personages.

---

1. i.e. the finished product is what one would expect of a good servant, even if his laziness is not. (This is Rodgers' interpretation of a cryptic sentence.)

The Poem

Pasiphilus, pearl of trust, to you I rightly
disclose the shrouded secrets of my heart.
Those fourteen little books, the *Work of Farming*,
penned by this hand with footed metre dumb ² –
unshaped by rhythm, untouched by Apollo's flow,   5
just rough-and-ready in pure rusticity –
you commend those mundane words, esteem and love
and cherish them as a warm supportive friend.
So now my growing confidence presents
this modest poem for you to judge and enjoy.   10

My Muse's not unpardonable aim
is to pen an urbane work of rustic ways;
to couple trees in a kind of fruitful marriage,
so a beauty blent from each grows in their young;
to clothe a wedded grove in allied leaves,   15
and exalt the offspring with twin foliage;
to mingle their sweet juices in glad union
and enjoy delightful fruit of dual flavour.
I shall teach which plants accommodate which others,
and which are laden with adopted leaves.   20

The very ruler of the cosmos, by whose will
the bright stars glide, earth stands, and seatides flow,
could have strewn a medley of flowers along the branches

---

2. In addition to the play on hand and foot, this line suggests a contrast between
silent writing of prose and the oral quality of verse.

and coloured the fruitful grove with varied leaves;
but he deigned to grant that honour to human toil   25
and sanctioned nature being made new by art.

I do not think my Muse's task is idle;
the credit for this small work will not be meagre.
If a swift and ardent mare is mated to a slow
donkey, so the union falls to a sterile level   30
and the heir produced undoes his fertile lineage
(procreative wealth engendering its own dearth),
may a poor tree not be enriched by buds it hosts
and shine by receiving the glory of a foreign flower?[3]
I shall make a start: what old-time farmers have written   35
I shall toil to follow – the sacred words of the ancients.

Firstly, persistence and skill have devised many types
of grafts, and guide the trained hand's intervention.
Any verdant tree adorned with another's leaves
has learnt to wear the offering in one of these ways: 40
either new scions are fastened behind the bark,
or the trunk is split to receive another top,
or an eye is replaced by a swelling foreign bud
bound by its gummy bark in that soft hollow.[4]

First to be grafted was the *vine* of Theban Bacchus,   45
and still its grapes swell up with an alien *cru*.
Embraced and nurtured by the adult plant,
the adopted buds enrich its twining limbs;
their well-bred foliage shades the inferior branches
and bows them low beneath the god's rich burden.   50

3. i.e. if stockbreeders produce infertile animals from fertile ones, it is surely
creditable for grafters to achieve the opposite result in plants.
4. The methods referred to are successively 'in the bark' and 'in the split trunk'
(for these cf. 3.17) and finally shield-budding (7.5.2–4); the 'hollow' is that left by
removal of the original eye from the host tree.

Wild trees are improved by branches of Pallas' *olive*,
whose noble berries enhance the feral fruit;
yet the barren wildling quickens the richer trees
and teaches yields of which it has no knowledge.[5]

The pallid *pear* lends out its snow-white flowers   55
with generous love, and weaves a dappled grove;
now strips her prickly sisters, the untamed pears,
of their fearsome weapons, teaching them to disarm;
now lengthens the rounded apple to a fertile tip
and bows the ash tree's hands with novel honours.   60
She even sets Phyllis producing fruit that is bigger,
softer, and smooths her harsh skin with her own.[6]
Barren thorns and the idle flowering-ash she dowers
with crops, and makes them love this unknown glory.
Her implanted branches have altered quinces too,   65
their mingled essence creating delightful fruit.
She strips the chestnut of its rough-coated yield,
changing its burden for a cultivated freight;
disarms the warlike limbs of the spiky medlar
and buries its fierceness beneath good-tempered bark.   70
They say she bonds her shoots with Libyan branches
and is able to take delight in Punic splendour.[7]

Punic apples have never fancied the flavours
of other fruits, or merged with disparate leaves.[8]
They enhance their own buds by exchanging genes,   75
and charm themselves by wearing the family crimson.

5. i.e. the cultivated olive is grafted onto wild olive stock, to their mutual benefit
(cf. 5.2.1–2).
6. Phyllis stands for the almond tree, into which she was metamorphosed
7. The Punic tree or Punic apple is the pomegranate, with its 'Libyan' i.e. North
African origin.
8. i.e. despite the claim that pear can be grafted onto pomegranate (verses 71–72),
pomegranate itself is never grafted onto different stock.

Grafted on vertical branches, the friendly *apple*
coalesces gladly, and improves her partner the pear.[9]
She also resolves to leave her own wildness behind
in the bush, and delights to savour a nobler harvest.[10]   80
Spine-bearing plums and thorn-bushes' armoured stems
she smooths and clothes with her growth of lovely leaves.
With her sweet juice she can swell the little sorb
to a size that bows the fruit to eager hands.
She delights to change the name of willow trees   85
and strew with flowers the grove beloved of the Nymphs.[11]
The plane, congenial to Bacchus the thyrsus-bearer,[12]
she teaches to redden, filled with a novel crop.
She provides the peach with unwonted greenery
to admire, and leafy poplars bear bright gifts.   90
For her the medlar changes its stony flesh,
expands and reddens, filled with snow-white juice.
In place of gravid spikes and pregnant armour,
chestnuts produce new apples, a golden glory.

With better shoots the *peachtree* loads the branches   95
of her kind, and deftly allies her genes with the plum;
she bestows her flimsy shade on Phyllis' trunk,
and learns to grow more vigorous by that move.

The golden *quince*, though she hosts all kinds of fruits,
does not lend herself as a guest to any tree else.   100
She proudly spurns the bark of an alien trunk,
knowing that none can grow her glory so well.
Spreading a kindred bed for lineal branches,
she stands content to enhance her own noble stock.

9. Pear trees have a characteristic upright habit of growth, hence 'vertical branches'.
10. i.e. cultivated varieties of apple are grafted onto wild apple stock.
11. Since the Nymphs inhabit trees and water, willows are doubly attractive to them.
12. Here and at line 124 planes are mentioned as support trees for vines, a role more usually assigned to poplar, elm and ash.

Supplanting the pear's hard fruit, scorned for its flavour,   105
the *medlar* feels secure once its scion is installed:
thus grafted it grows even fiercer, doubly-armoured,[13]
and scares off greedy hands with its fearsome limbs.

The *citron's* branches allow the loan of their offspring,
which the mulberry nurtures in its teeming bark;[14]   110
and to nourish fruits scented with delightful juice,
they amend the armoured pear tree's well-known darts.

Plum trees bestow rich shoots on plum tree limbs,
and garner fertile gifts from cognate bodies.
When bidden to make its home in chestnut trees,
the plum disarms their crop but arms their boughs.   115
Carobs grow used to softening with green juice,
and nurture in their bosom waxy fruits.

Mulberries are coaxed to vary their awful black
by the *fig*, imposing its norms on occupied trees.   120
In turn it swells (astonished) with rich juice,
and is thrilled by fruit outdoing its normal size.
The plane has signal leaves, feast-shading limbs,
a frame that delights in the glory of the vine:
the incoming fig, sustained by its own rich bark,   125
flourishes, grips and fills the desirable hollows.[15]

In fact *black mulberries* requite the exchange with the fig,
which nurtures shoots that the mulberry provides.
The ash presents its limbs to this keen companion,

13. Like other grafts, this one is made onto wild stock, here the wild pear: hence the
reference to hard fruit, and to the wild pear's armour (cf. ll. 57–58) reinforcing that
of the medlar itself.
14. These lines refer to reciprocal grafting from the citron to the black mulberry and
back again (*Geoponika* 10.7.12). The language is also skilfully reciprocal, since 109
could be understood as 'the citron's branches receive shared offspring.'
15. Again a reference to shield-budding, for which the fig with its rich bark is well
suited; the hollows (*sinus*) are those left in the bark of the host tree by removal of
'shields'.

grows blood-bespattered, and fears its novel brood.   130
High beeches, and the vigorous chestnut's fruit
(bristling and rough amidst harsh leaves) – the mulberry
stains these, shows how to darken with pitchy young,
and feeds the swelling fruit with novel juice.
The sweetly-scented terebinth complies,   135
and blended gifts result, with the merits of each.

Sorbs augment their yield with better stock
of their own, and glisten, bowed by a splendid burden.
This tree removes the spikes from the doughty limbs
of the thorn, and hides its weapons under placid bark;   140
she mingles her grafted crop with golden quinces
gladly, and loves the results with their borrowed hues.

The *cherry* is inserted in the virgin laurel, who reddens
with a borrowed blush from the children she is forced to bear.[16]
Shoots of cherry impose their colour on the boughs   145
of shade-giving planes and plum trees vicious-stemmed,
and fleck the branches of poplar with novel gifts,
so much does their sweet red sprinkle those grey arms.

Phyllis, installed in the plum's divided bark,
covers its fragrant limbs with early flowers.[17]   150
She changes peaches into fruit with hybrid coats,
and trains her hard shell to replace their skin.
She curbs and rounds the carob's lengthy shape,
enriches its uncouth leaves with lovely fragrance;
makes the fierce chestnut drop its sea-urchin spikes   155
and admire the smoothness of the fruit it bears.

16. The virgin is Diana, who was transformed into a laurel when pursued by Apollo.
17. Again Phyllis represents the almond tree.

In turn *pistachios* supersede almond branches,
their goal being 'greater quality by smallness!'
The terebinth too enfolds them in kindred dress,
and feeds them, hoping to be bettered by their adoption.   160

The lofty limbs of the *chestnut* fecundate
the riverside willow, and thrive on so much moisture.

The mighty *walnut* invades the arbutus' boughs,
and renders fruit protected by a double shell.

Other grafts of novel types have been invented,   165
which sage experience will unfold in time's advance.
Those named above will suffice this modest poet,
who delights to quicken the surface of a well-dug field.
These verses you read were fashioned amidst hard hoes –
rough verses, but with a mellow country flavour.   170

FIGURES

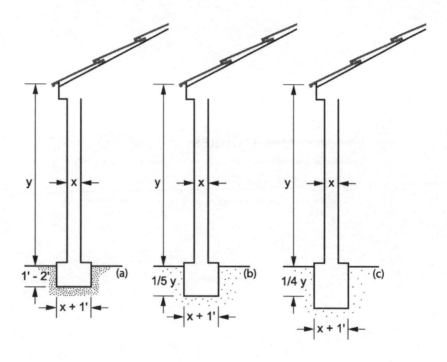

Figure 1. Foundations for villa walls (Palladius 1.8): (a) on rock; (b) on solid clay; (c) on loose ground. After Plommer 1973: 17.

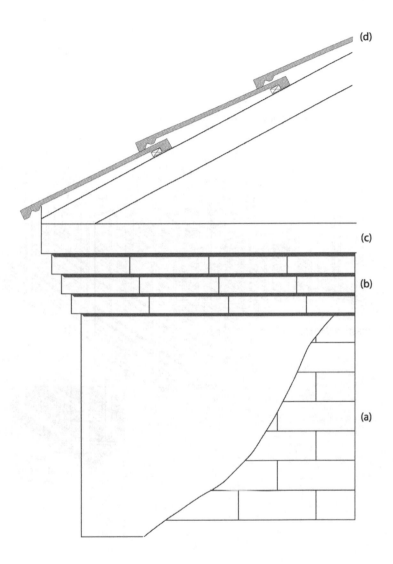

Figure 2. Protection for brick walls (Palladius 1.11): (a) brick wall with plastered surface; (b) tile coping; (c) wooden beam; (d) roof tiles (capping tiles not shown). After Martin 1976: 123.

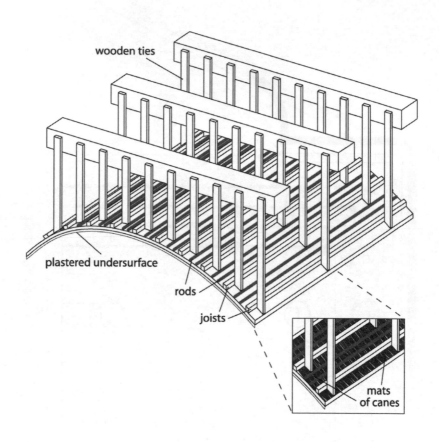

wooden ties

plastered undersurface

rods

joists

mats
of canes

Figure 3. Suspended ceiling using canes (Palladius 1.13). After Howe
(Rowland 1999: 271).

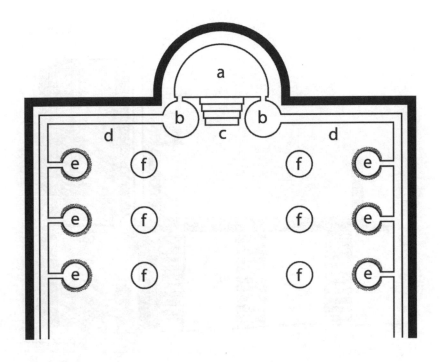

Figure 4. Plan of wine-pressing room (Palladius 1.18): (a) raised pressing floor; (b) vats; (c) steps; (d) channels or pipes; (e) pots; (f) casks. After Hamblenne 1980: 168.

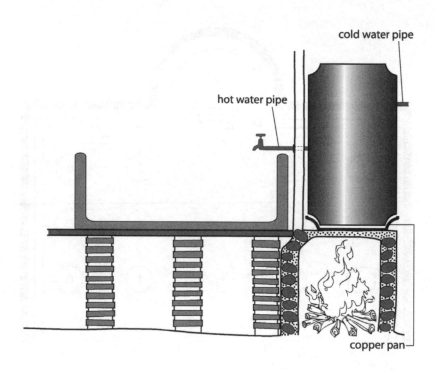

Figure 5. Heating system for baths (Palladius 1.39.3).

Figure 6. Gallo-Roman reaping machine (Palladius 7.2.2–4).

Appendix:
Divergences from Rodgers' Latin Text

My translation reflects the following textual differences from the Teubner edition of Palladius edited by R.H. Rodgers. Rodgers made a few adjustments to his text in his *Introduction to Palladius*. The arguments for my conjectural emendations are given in my article in *HSCP* vol. 107.

	Rodgers	Fitch
1.5.3	*iudicio*	*indicio* (Marshall)
1.5.4	*coloris et corporis*	[*coloris et*] *corporis* (Schneider)
1.8.3	*hiemali*	*aestiuo* (Fitch)
1.17	title	deleted (Fitch)[1]
1.30.4	*eorum*	*haedorum* (Pontedera)
1.35.2	*culices*	*pulices* (Fitch)
1.35.3	*propter multa portenta*	*propter multam potentiam* (Fitch)
1.35.4	*cantharides*	*campas* (Fitch from *Geoponika*)
1.37.8	*uitentur*	*utentur* (early editions)
1.40.1	*calidarum*	*calidariarum* (Widstrand)
1.40.2	*super*	*et sulphur* (Schneider from Faventinus)
2.2	*apricis aut*	*apricis et* (from Columella)
2.13.3	*subposita*	*superposita* (Aldine from Columella)
2.13.3	*saturabit*	*saturauit (M)*

1. The title of 1.17 is interpolated. I have replaced it with the title of 1.18, and combined the two chapters into one.

	Rodgers	**Fitch**
2.17	no lacuna	lacuna after *hoc modo* (Fitch)
3.9.2	*atque fecundas*	*minusque fecundas* (Fitch)
3.12.1	*ne debili*	*nec debili* (Fitch)
3.12.2	*optimum putaturis*	*optimi* (M) *putatoris* (MSS)
3.13.2	*primi anni*	*proximi anni* (from Columella)
3.13.3	*infra*	*intra* (Fitch)
3.17.8	*infra*	*intra* (Schneider)
3.25.21	*largae arbores*	*[largae] arbores* (Fitch)
3.25.26	*clauduntur*	*claudunt* (M)
3.25.30	*terebinthi, hinc inde*	*terebinthi hinc, inde*
3.25.32	*inseretur*	*inseritur* (Rodgers *Introduction*)
3.25.32	*ossa prunorum*	*<ponuntur> ossa prunorum* (Fitch)
3.31.2	*et locis*	*ex locis* (MSS)
4.9.11	*est stercore*	*[est] stercore* (Widstrand)
4.10.1	*uelut*	*uelut <sarmentum>* (Fitch)
4.10.6	*qui est*	*qui inest* (Aldus)
4.10.7	*post triduum*	*per triduum* (Fitch)
4.10.32	*propagari ficus ramis potest.*	deleted (Fitch)
4.10.33	*alii missas*	*alii [missas]* (Fitch)
4.10.35	*toto sole*	*toto <die> sole* (from Columella)
4.11.2	*quadratis et grandibus*	*quadrati, grandibus* (Schoettgen from Columella)
4.11.4	*quibus cordi est*	*cui cordi est* (Martin)
4.11.4	*aut his signis*	*his signis aut* (Martin)
4.11.5	*alta fronte*	*lata fronte* (Crescentius)
4.13.8	*infra sextum*	*intra sextum* (from Columella MSS)

	Rodgers	Fitch
6.7.3	*sagina*	*saliua* (Fitch)
6.7.4	*strictis*	deleted (Schneider)
7.3.1	*occabimus. colligemus*	*occabimus, <seminaria fodiemus>, colligemus* (Fitch)
7.5.4	*adpositae*	*adposita* (G)
7.11	*infuso*	*infusum* (Fitch)
8.2.1	*locis umidis*	*<rapa> locis umidis* (Schneider)
9.12	*nongentas*	*octingentas* (MSS)
10.1.1	*primo Augusto*	*primum Augusto* (from Columella)
11.14.13	*item uinum quod fieri enim*	*item quod* (DPKJLG) *fieri mustum* (Fitch)
12.7.13	*perferuntur*	*proferuntur* (GS)
12.7.14	*laetamine, excusant*	*laetamine excusante* (Crescentius)
12.7.16	*haec sunt ... dicuntur*	sentence deleted (Fitch)
12.7.20	*fidentibus*	*findentibus* (MSS)
12.22.4	*supra conpositas bacas refundes*	deleted (Fitch)
13.4.2	no lacuna	lacuna at end (Fitch)
14.3.2	*myrti*	*<folia> myrti* (Svennung)
14.4.4	*potionibus*	*portionibus* (Corsetti)
14.5.3	*tunc panacis <et> eryngii radices*	*tympanicis eringi radices* (MSS)
14.6.1	*brassicae caules*	*brassicae <modicae> caules* (MSS of Columella)
14.6.7	*salis sextarius*	*salis sextans* (from Columella)
14.6.7	*uel [salis] admixto*	*et [salis] admixto* (from Columella)

	Rodgers	**Fitch**
14.7.2	quindecim coni totidemque gallae	quindecim pondo idemque gallae (MSS)
14.8.1	perluitur	perluitur <os> (Svennung from Columella)
14.13.9	faeces	fraces (Corsetti from Columella)
14.14.4	iungitur	iniungitur (from Columella)
14.14.7	utendum est et cum	utendum est [et]: cum (Svennung from Columella)
14.14.10	priusquam adiungantur	priusquam diiungantur (Corsetti from Columella)
14.15.2	altius ferro	latius ferro (from Columella)
14.15.5	mulsa frigida	multa frigida (Columella, v)
14.19.3	adiciunt	adiciuntur (from Columella)
14.20.2	nidor deusti cimicis <inmitti	[nidor] deusti cimicis <odor inmitti (Richter from Columella)
14.22.8	adaeque	equae Richter (equo Columella)
14.22.8	fouendae sunt	fouendae <et axungia uetere confricandae> sunt (from Columella)
14.22.10	per triduum	post triduum (from Columella)
14.24.6	ex aeno nouo	ex oleo nouo (MSS)
14.25.1	insectantes	infestantes (Richter from Columella)
14.25.4	leuatur	liberatur (MSS)
14.28.5	est hyosciamos	est <et> hyosciamos (from Columella)
14.28.8	uinumque suffunditur	uinumque <in os> suffunditur (Richter from Columella)

APPENDIX

	Rodgers	Fitch
14.30.1	spinis sauciari	spinis <ac rubis> sauciari (Richter from Columella)
14.30.3	primo tempore	primo <quoque> tempore (Richter from Columella)
14.30.6	promitur	promptum (MSS)
14.32.6	Bolus	ciuis (MSS)
14.32.8	ueteris cunelae	[ueteris] cunelae (from Columella)
14.35.4	iam cytisi	etiam cytisi (Sabbadini) (καὶ κυτίσσου Geoponika)
14.36.2	ita etiam	item (from Columella)
14.36.7	permixta	<trita et> permixta (from Columella)
14.39.1	pabulo	pabula (MS M)
14.50.1	non frigescat	ne frigescat (Fitch)
14.50.1	et postea sanguinem tollis	deleted (Fitch)
14.50.2	aut lactis	[aut] lactis (Fitch)
14.59	cataplasma proficiet	cataplasmans (v) proficies (Mv)
Poem l.18	laetum duplici	laeta duplicis (Fitch)
Poem l.44	accipit	excipit (Fitch)

Select Bibliography

Editions and translations of Palladius

Ana Moure Casas (transl.), *Paladio: Tratado de Agricultura*. Madrid, 1990.

J.M. Gesner, ed., *Scriptores rei rusticae veteres latini*, vol. 2. Leipzig, 1735.

C. Guirard, ed., and René Martin, transl., *Palladius: Traité d'Agriculture*, tome II (Livres III à V). Paris: Les Belle Lettres, 2010.

John Henderson, *HORTVS: The Roman Book of Gardening*. London: Routledge, 2004. (Includes translations of Palladius' chapters on gardens.)

Mark Liddell, ed., *The Middle-English Translation of Palladius De re rustica*. Berlin, 1896.

René Martin, ed. & transl., *Palladius: Traité d'Agriculture*, tome I (Livres I et II). Paris: Les Belle Lettres, 1976.

T. Owen, transl., *The Fourteen Books of Palladius Rutilius Taurus Aemilianus on Agriculture*. London, 1807.

R.H. Rodgers, ed., *Palladii Rutilii Tauri Aemiliani Opus agriculturae, De veterinaria medicina, De insitione*. Leipzig: Teubner, 1975.

C.F. Saboureux de la Bonnetrie, *Traduction d'anciens ouvrages latins relatifs à l'agriculture et à la medicine vétérinaire*, tome 5. Paris, 1775. (Reprinted with revisions as *Les Agronomes latins, publiés sous la direction de M. Nisard*. Paris, 1844.)

J.G. Schneider, ed., *Scriptorum rei rusticae veterum latinorum tomus tertius, Palladii Rutilii Tauri Aemiliani De re rustica libros XIV tenens*. Leipzig, 1795.

Books and articles

J.N. Adams, *Pelagonius and Latin Veterinary Terminology in the Roman Empire*. Leiden, 1995.

Mauro Ambrosoli, *The Wild and the Sown: Botany and Agriculture in Western Europe 1350–1850*. Cambridge, 1997.

Jacques André, *Les Noms des plantes dans la Rome antique*. Paris, 1985.

Marco Johannes Bartoldus, *Palladius Rutilius Taurus Aemilianus: Welt und Wert spätrömischer Landwirtschaft*. Augsburg, 2012.

J.-P. Brun, *Archéologie du vin et de l'huile dans l'empire romain*. Paris, 2004.

Pierre-Paul Corsetti, 'A propos d'une édition récente de Palladius. Remarques sur la tradition manuscrite et le texte du livre XIV', *Latomus* 37 (1978): 726–746.

Andrew Dalby, *Food in the Ancient World from A to Z*. London, 2003.

A. De Angelis, *La coltivazione delle piante da frutto nella letteratura agronomica latina*. Rome, 1995.

John G. Fitch, 'Textual Notes on Palladius, *Opus Agriculturae*', *Harvard Studies in Classical Philology* 107 (forthcoming).

G. Forni, A. Marcone (eds.), *Storia dell'agricoltura italiana* I.2: *L'età romana*. Florence, 2002.

Cam Grey, 'Revisiting the 'problem' of agri deserti in the Late Roman Empire', *Journal of Roman Archaeology* 20 (2007): 362–76.

Pierre Hamblenne, 'Réflections sur le livre Ier de l'*Opus agriculturae* de Palladius', *Latomus* 39 (1980): 165–172.

W. Kaltenstadler, 'Arbeits- und Führungskräfte im *Opus Agriculturae* von Palladius', *Klio* 66 (1984): 223–229.

Geoffrey Kron, 'Animal Husbandry, Hunting, Fishing, and Fish Production', in Oleson: 175–222.

Evi Margaritis and Martin K. Jones, 'Greek and Roman Agriculture', in Oleson: 158–74.

Frank Morgenstern, 'Die Auswertung des *Opus agriculturae* des Palladius zu einigen Fragen der spätantiken Wirtschaftsgeschichte', *Klio* 71 (1989): 179–192.

John Peter Oleson, *Engineeering and Technology in the Classical World*. Oxford, 2008.

J. Peters, *Römische Tierhaltung und Tierzucht. Eine Synthese aus archäozoologischer Untersuchung und schriftlich-bildlicher Überlieferung.* Rahden, 1998.

H.W. Pleket, 'Die Landwirtschaft in der römischen Kaiserzeit', in F. Wittinghoff (ed.), *Europäische Wirtschafts- und Sozialgeschichte in der Römischen Kaiserzeit*, I. Stuttgart, 1990.

Hugh Plommer, *see under* Cetius Faventinus.

Will Richter, 'Palladius und sein Columella-Text im Buch über die Tiermedizin', *Wiener Studien* 12 (1978): 249–71.

R.H. Rodgers, *An Introduction to Palladius.* London 1975 [a detailed study of the manuscripts and textual transmission].

J.J. Rossiter, *Roman Farm Buildings in Italy.* Oxford, 1972.

J. Svennung, 'De Auctoribus Palladii', *Eranos* 25 (1927): 123–78, 230–48.

——, *Untersuchungen zu Palladius und zur lateinischen Fach- und Volkssprache.* Uppsala, 1935.

David L. Thurmond, *A Handbook of Food Processing in Classical Rome: For her Bounty no Winter.* Leiden, 2006.

Domenico Vera, 'I silenzi di Palladio e l'Italia: osservazioni sull'ultimo agronomo romano', *Antiquité Tardive* 7 (1999): 283–297.

K.D. White, *Agricultural Implements of the Roman World.* Cambridge, 1967.

——, *Roman Farming.* London, 1970.

——, *Farm Equipment of the Roman World.* Cambridge, 1975.

Editions and translations of related texts

Cetius Faventinus

Marie-Thérèse Cam, ed. & transl., *Cetius Faventinus: Abrégé d'Architecture Privée*. Paris: Les Belles Lettres, 2002.

Hugh Plommer, *Vitruvius and Later Roman Building Manuals*. Cambridge, 1973.

Columella

Harrison Boyd Ash, E.S. Forster, Edward H. Hefner, transll., *Columella: On Agriculture*. 3 vols. Cambridge, MA: Harvard University Press, 1941–55.

R.H. Rodgers, ed., *L. Iuni Moderati Columellae Res Rustica*. Oxford, 2010.

Geoponika

Heinrich Beckh, ed., *Geoponica sive Cassianai Bassi scholastici de re rustica eclogae*. Leipzig: Teubner, 1895.

Andrew Dalby, transl., *Geoponika: Farm Work*. Prospect Books, 2011.

Emanuele Lelli, transl., comm., *L'Agricoltura Antica: I Geoponica di Cassiano Basso*. 2 vols. Soveria Mannelli: Rubbettino, 2010.

Gargilius Martialis

A. Mai, ed., *Gargilii Martialis de arboribus pomiferis fragmentum ex codice rescripto bibliothecae regiae Neapolitanae*, in *Classicorum auctorum e Vaticanis codicibus editorum*, tom. 1. Rome, 1828: 387–413.

S. Condorelli, ed., *Gargilii Martialis Quae Exstant*. Rome, 1978.

Index

abineus, 140

ablaqueation, 29, 153; of almond trees, 81; of fig trees, 134; of mulberry trees, 114; of olive trees, 188, 189; of palm trees, 191; of peach trees, 204; of pear trees, 108; of pomegranate trees, 128; of quince trees, 112; of vines, 71, 122, 152, 187, 202

abscesses, in cattle, 231; in horses, 243

acinaticium, 40

acorns, 115, 212

aesculus oak (*Quercus frainetto*), for building, 213; for flooring, 45

Africa, North, 199

African pea meal, 196

aged urine, 122, 128, 134, 189; bovine, 224, 229; human, 123, 132, 207, 210

ailments, apple trees, 110–111; pear trees, 108; quince trees, 112

air, assessment of, 36

Albertus Magnus, 24, 25

alder wood, 213

alfalfa, 87, 144, 180, 211, 223, 244

almond trees, 65, 79-81, 108, 114, 192–193, 205, 207, 267

almonds with writing on them, 81, 168

almonds, bitter, 195, 256

aloes, 195; hepatic, 196, 197

alum, 224, 241, 247, 257; split alum, 224, 233, 241

Aminean vines, 88

Aminean wine, 142

ammoniac gum, in caulking, 69

ammoniac rock salt, 54, 224, 238

amurca, 29, 40, 51, 61, 62–64, 87, 94, 112, 118, 123, 128, 132, 134, 155, 159, 179, 189, 205, 207, 214, 224, 226, 233, 246, 246, 250, 254

Anatolius of Beirut, 13

anise, 107, 127

ant eggs, 56

antitoxin vine, 116–117

antitoxin wines, 221

ants, 61, 132, 134, 159, 193, 207

aphrodisiacs, 140, 254–255, 257

Apian vines, 88

apple trees, 108, 110–111, 132, 136, 168, 183, 207; ailments, 110–111; grafting, 111, 265; planting, 193; propagation, 110; worms in, 110

apple vinegar, 111

apple wine, 111

apples, 161; preserving, 111

April, tasks, 144–150

Apuleius, 63

aqueducts, 176–177

Arabia, 250

arbutus, 54, 66, 83, 100, 212, 268

argil, 59, 76, 104, 110, 111, 113, 129, 132, 178

Aristotle, 169

neck ailments, in cattle, 234
'needles', 70
nettles, 57, 222, 242
nigella, see git
nightshade, black, 223, 255–257; stinking,
193, 223, 257
nitrum, 105, 202, 214, 224, 241, 257
North Africa, 199
Norton, Thomas, 23, 24
nose-bleeds, in horses, 242
November, tasks, 201–217
nursery beds, 91–92, 101, 102, 103, 123, 145,
153, 160
nuts, to smoke out moles, 63

oak, 45, 52, 65, 108, 123, 213; *aesculus* oak
(*Quercus frainetto*), 45, 213; *cerrus* oak
(*Quercus cerris*), 45, 100; cork oak, 66;
holm-oak, 100; oak ash, 63, 123
October, tasks, 186–200
oil, 83–84; camomile-flavoured, 165 laurel,
203, 218; lily-flavoured, 157; mastic,
218, 225, 238, 242; myrtle, 218; rose-
flavoured, 157, 256; Spanish, 171; see
also olive oil
oil room, construction of, 51–52
oleander leaves, 51, 63
olive oil, 51–52, 125, 134, 148, 156, 157, 203,
212, 214–215, 228, 229, 231–240, 242,
244, 248, 255, 257, 258; cleaning dirty,
214; curing rancid, 215; foul-smelling,
214–215; lees, 233; Liburnian, 214;
made from green olives, 189-190, 214;
old, 234, violet-infused, 148
olive trees, 39, 40, 41, 48, 264; February,
99-100; March, 123; October, 188–189;
ablaqueation of, 188; grafting, 144–145;
leaves, 51, 230; manuring, 188–189;
planting, 99-100, 203; propagation,
42; pruning, 203; shield-budding, 161;
sterile, 203; varieties, 100; wild, 207,
223, 228, 229, 264
olives, black, 216; parti-coloured, oil
from, 203; preserving, 188, 215–216;
seasoning, 215–216; 'swimming', 215;
varieties, 100; white, 188; wild, 54

onions, 82–83, 104, 124, 135, 146, 167, 191,
204, 218, 225, 228, 254, 257
orach, 146, 167
orchards, 59
orchis olive, 100
oregano, 65, 66, 124, 182, 191, 216, 217,
223, 255
orf, in lambs, 249
ostigo, 249
owls, 61; snares to catch, 182
ox, oxen, bile, 62; buying, 136–137; dung,
82, 143, 209; gall, 63, 110, 128; manure,
108, 110, 183; marrow, 225; tallow, 224;
taming, 138–139; urine, 64, 207
oyster shells, 69

paguri, 62
Palladius Rutilius Taurus Aemilianus,
life and identity, 11–12; audience
for *Work of Farming*, 17-19; Book
14 on veterinary medicine, 19–20;
composition of *Work of Farming*,
13–15; style of *Poem on Grafting*, 20;
style of *Work of Farming*, 15–16, 262
palm trees, date palms, 147, 153, 161, 168;
ablaqueation of, 191; baskets, 116, 130,
198; planting, 191
papyrus, 117, 119, 222, 228
parsley, 146, 153, 161, 164; horse-, 146;
marsh-, 146; rock-, 146, 216; seed, 257
Pasiphilus, 20, 261ff
pasture, 172, 181, 202, 211
pausian olive, 100
pavements, paving, 50–51, 156, 165, 184
pea, African, meal, 196
peach trees, 65, 83, 114, 153, 161, 204–205,
207, 265, 267; ablaqueation of, 204;
Armenian, 83, 205; grafting, 99, 147,
205, 265; varieties, 205
peaches, grown with writing on them,
204; preservation of, 205
peafowl, 55, 56
pear, pears, 161; preserving, 109; sauce, 110;
vinegar, 109; wine, 109
pear trees, 65, 131, 132, 136, 168, 183, 264;
ailments, 108; February, 107–110;

INDEX

sudden wasting, in horses, 243
sulphur, 62, 63, 69, 224, 233, 235, 241, 242, 244, 246, 247, 250, 254, 256
sumach, Syrian, 142
summer savory, 105, 124, 204, 218, 222, 230, 248
suppurated wound, in cattle, 254
suspended ceilings, 47–48, fig. 3
Svennung, Josef, 19
swarms, bee, 66–67, 142, 148–149, 156, 163–165
swellings, in horses, 257
swimming, of benefit for horses, 258
'swimming' olives, 215
swollen glands, in pigs, 251
Syrian sumach, 142
Syrians, references to, 195

tail, stag's, 254–255; wolf's, 253
tallow, 69, 247; bovine, 231; goat, 231, 232, 234; goat or ox, 224
tamarisk, honey, 209; wood, 252
Tarentine almonds, 82
Tarentine sheep, 212
teeth, horses, 141
tenants, 39, 245
terebinth, 65, 134, 192, 268; resin, 257; wood, 205
terra rossa, 156, 180, 208
text, transmission errors in, 27, 45n, 83n, 154n, 164n, 189n, 219n, 227n, 228n, 229n, 230n, 241n, 248nn, 277–81
thapsia, 66
third-ploughing, 178
thistle flowers, for rennet, 155
thorn, thorns, 58, 245; hedges, for gardens, 59–60, 103; trees, 108, 111, 114, 183; see also whitethorn
three-month crops, 42
threshing-floor, 59, 64–65, 159
thrushes, 53–54, 220
thymbra, 66
thyme, 65, 155, 182, 191; creeping, 65, 66, 127, 146, 222; honey, 225, 242
timber, 85, 153, 212–213, 218
titymallus, 222, 235

tools, 70
torch, pine, 223
tortoise shell, 62
tow, in caulking, 69
training vines, 151
transplanting, 79, 80, 82, 89, 91–93, 99, 101, 105, 106, 114, 117, 126, 130, 133, 146, 153, 183, 188, 190–192, 204, 206–208, 210, 213, 218, 219
tree fruits, 83, 114, 136, 147, 161, 191, 193, 209; see also under specific varieties
tree-medick, 251, 254
trees, 76, 91, 92, 95, 97, 100, 102, 115, 152, 167, 174, 212–213; see also fruit trees, and specific varieties
trefoil, meadow, 237; mountain, 222, 237; pointed (?*Psoralea bituminosa*), 237
trimming, vines, 151–152, 172
truncheon, 130, 131
tuberculosis, in cattle, 233
tufa, 36, 44, 76, 113, 175, 208
tunics, 70
Turkey oak (*Quercus cerris*), 45, 100
turnips, 62, 167, 168, 173, 181, 183; preserved, 83; seasoning, 219
turtle, marsh, 64; sea, blood, 224
turtle-doves, 53

ulcerated lungs, in cattle, 233
ulpicum, 78, 104, 124, 203–204, 218, 225, 226
unripe-grape juice with honey, 177
urine, 242; aged, 122, 128, 134, 189, 250, 253, 254; aged bovine, 224, 229, 254; aged human, 132, 207, 210, 231, 246, 247, 248; as fertilizer, 87; ass, 129; blood in the, in horses, 257; bovine, 231; donkey, 250; human, 110, 257; ox, 64, 207; stoppage of, 239, 255, 257

variegated grape clusters, 119
Varro, 25
vats, for wine, 50
venom, antidotes for, 221
verdigris, 225, 247, 249
Vergil, 19, 20, 22, 25, 108, 246